SpringerBriefs in Mathematics

SpringerBriefs present concise summaries of cutting-edge research and practical applications across a wide spectrum of fields. Featuring compact volumes of 50 to 125 pages, the series covers a range of content from professional to academic. Briefs are characterized by fast, global electronic dissemination, standard publishing contracts, standardized manuscript preparation and formatting guidelines, and expedited production schedules.

Typical topics might include:

A timely report of state-of-the art techniques A bridge between new research results, as published in journal articles, and a contextual literature review A snapshot of a hot or emerging topic An in-depth case study A presentation of core concepts that students must understand in order to make independent contributions

SpringerBriefs in Mathematics showcases expositions in all areas of mathematics and applied mathematics. Manuscripts presenting new results or a single new result in a classical field, new field, or an emerging topic, applications, or bridges between new results and already published works, are encouraged. The series is intended for mathematicians and applied mathematicians. All works are peer-reviewed to meet the highest standards of scientific literature.

Titles from this series are indexed by Scopus, Web of Science, Mathematical Reviews, and zbMATH.

More information about this series at http://www.springer.com/series/10030

Atsushi Yagi

Abstract Parabolic Evolution Equations and Łojasiewicz–Simon Inequality II

Applications

Springer

Atsushi Yagi
Osaka University
Suita, Osaka
Japan

ISSN 2191-8198   ISSN 2191-8201   (electronic)
SpringerBriefs in Mathematics
ISBN 978-981-16-2662-3   ISBN 978-981-16-2663-0   (eBook)
https://doi.org/10.1007/978-981-16-2663-0

This Springer imprint is published by the registered company Springer Nature Singapore Pte Ltd.
The registered company address is: 152 Beach Road, #21-01/04 Gateway East, Singapore 189721,
Singapore

Preface

This second volume continues the study on asymptotic convergence of global solutions of parabolic equations to stationary solutions by utilizing the theory of abstract parabolic evolution equations and the Łojasiewicz–Simon gradient inequality. In the first volume [Yag], after setting the abstract frameworks of arguments, a general convergence theorem has been proved under the four structural assumptions, Critical Condition, Lyapunov Function, Angle Condition and Gradient Inequality. In this volume, after reviewing those abstract results briefly, we want to describe their applications to some concrete parabolic equations.

In Chap. 3, we treat semilinear parabolic equations of second order in general n-dimensional spaces. Chapter 4 is devoted to treating epitaxial growth equations of fourth-order which incorporate general roughening functions. Chapter 5 is devoted to considering the Keller–Segel equations in one-, two- and three-dimensional spaces. Some of these results were already obtained in the papers Azizi–Mola–Yagi [AMY17], Grasselli–Mola–Yagi [GMY], Iwasaki–Osaki–Yagi [IOY], Iwasaki–Yagi [IY]. However, by means of the abstract theory described in the first volume, those results can be extended much more.

To read this monograph, knowledge of Functional Analysis of standard level ([Bre11, Yos80, Zei88] and so on) and the knowledge of Function Spaces ([Ada75, Gri85, Tri78] and so on) are required. The readers are also expected to be familiar with the functional analytic methods for partial differential equations ([DL84a, DL84b, Tan75, Tan97, Yag10] and so on).

Lectures on topics related to these two volumes have been given by the author in the postgraduate course of the Vietnam Institute for Advanced Study in Mathematics, Osaka University and Kwansei Gakuin University.

I would like to express my profound gratitude to Professors Nguyen Huu Du, Pham Ky Anh, Vladimir Fedorov, Gianluca Mola and Koichi Osaki for valuable comments and suggestions. I would also like to express my hearty thanks to my past students, Drs. Takeshi Uchitane, Nguyen Thi Hoai Linh, Ta Viet Ton, Somayyeh Azizi, Kensuke Ohtake, Maya Kageyama, Jian Yang and Satoru Iwasaki. The joint work with them has been important in preparing the two volumes.

Osaka, Japan   Atsushi Yagi
January 2021

Contents

Chapter 1
Preliminaries

In this monograph, we mainly handle real Banach spaces and Hilbert spaces and real linear operators. For this reason, Banach spaces, Hilbert spaces and linear operators always mean real ones if they are not otherwise specified. However, they are all obtained as a real part (in a certain sense) of some complex Banach spaces, Hilbert spaces or complex linear operators. This means that we can fortunately utilize the powerful tools developed in Complex Functional Analysis. The first half of this chapter is devoted to reviewing the most convenient way to utilize the tools of Complex Functional Analysis to real spaces and operators. These results were originally presented in the paper [Yag17].

The second half of this chapter is devoted to reviewing some results concerning Fréchet and Gâteaux differentiability for nonlinear operators from a Banach space into another.

This chapter concludes by recalling the definition of Fredholm operators and their important class which is given by Riesz–Schauder theory and by recalling the Łojasiewicz gradient inequality for analytic functions.

1.1 Basic Materials in Complex Functional Analysis

Let $X_{\mathbb{C}}$ and $Y_{\mathbb{C}}$ be two Banach spaces with norm $\| \cdot \|_X$ and $\| \cdot \|_Y$ respectively over the complex number $\mathbb{C}$. By $\mathcal{L}(X_{\mathbb{C}}, Y_{\mathbb{C}})$ we denote the space of all bounded linear operators from $X_{\mathbb{C}}$ into $Y_{\mathbb{C}}$. For each $T \in \mathcal{L}(X_{\mathbb{C}}, Y_{\mathbb{C}})$, its uniform operator norm is defined by $\|T\|_{\mathcal{L}(X,Y)} = \sup_{\|f\|_X \leq 1} \|Tf\|_Y$. Then, $\mathcal{L}(X_{\mathbb{C}}, Y_{\mathbb{C}})$ becomes a complex Banach space. When $X_{\mathbb{C}} = Y_{\mathbb{C}}$, $\mathcal{L}(X_{\mathbb{C}}, X_{\mathbb{C}})$ is written as $\mathcal{L}(X_{\mathbb{C}})$ for short. When $Y_{\mathbb{C}} = \mathbb{C}$, the bounded operator $\varphi \in \mathcal{L}(X_{\mathbb{C}}, \mathbb{C})$ is called a bounded linear functional on $X_{\mathbb{C}}$.

© The Author(s), under exclusive license to Springer Nature Singapore Pte Ltd. 2021
A. Yagi, *Abstract Parabolic Evolution Equations and Łojasiewicz–Simon Inequality II*,
SpringerBriefs in Mathematics, https://doi.org/10.1007/978-981-16-2663-0_1

1.1.1 Dual Spaces

In $\mathcal{L}(X_{\mathbb{C}}, \mathbb{C})$, we prefer to define a multiplication between the complex number α and the bounded linear functional φ as

$$[\alpha\varphi](f) = \overline{\alpha}\varphi(f) \qquad \text{for } f \in X_{\mathbb{C}}, \tag{1.1}$$

rather than to define $[\alpha\varphi](f) = \alpha\varphi(f)$, $f \in X_{\mathbb{C}}$, as usual. When $\mathcal{L}(X_{\mathbb{C}}, \mathbb{C})$ is equipped with this multiplication between the complex numbers and the bounded linear functionals, the space is denoted by $X'_{\mathbb{C}}$ which is a complex linear space. Furthermore, by the uniform operator norm mentioned above, $X'_{\mathbb{C}}$ also becomes a Banach space which we call *the dual space* of $X_{\mathbb{C}}$.

The definition (1.1) sometimes provides us an advantage over the usual one. Let $X_{\mathbb{C}}$ be a complex Hilbert space with inner product $(\cdot, \cdot)_X$. Then, for each $g \in X_{\mathbb{C}}$, the linear functional $(\cdot, g)_X$ is continuous on $X_{\mathbb{C}}$; therefore, a correspondence $J : g \mapsto Jg$ from $X_{\mathbb{C}}$ into $\mathcal{L}(X_{\mathbb{C}}, \mathbb{C})$ is defined by the relation

$$[Jg](f) = (f, g)_X \qquad \text{for all } f \in X_{\mathbb{C}}. \tag{1.2}$$

By virtue of the Riesz representation theorem, $J : X_{\mathbb{C}} \to \mathcal{L}(X_{\mathbb{C}}, \mathbb{C})$ is a bijection and an isometry, but J is conjugate linear, i.e., $J(\alpha g) = \overline{\alpha} Jg$. However, we verify that J is *linear* from $X_{\mathbb{C}}$ onto $X'_{\mathbb{C}}$ owing to (1.1).

Theorem 1.1 *If $X_{\mathbb{C}}$ is a Hilbert space, then the mapping $J : X_{\mathbb{C}} \to X'_{\mathbb{C}}$ defined by (1.2) is an isometric isomorphism.*

1.1.2 Adjoint Spaces

Let $X_{\mathbb{C}}$ and $Y_{\mathbb{C}}$ be complex Banach spaces with norms $\|\cdot\|_X$ and $\|\cdot\|_Y$ respectively. A complex-valued function $\langle \cdot, \cdot \rangle$ defined on the product space $X_{\mathbb{C}} \times Y_{\mathbb{C}}$ having the properties

$$\begin{cases} \langle \alpha f + \beta \tilde{f}, g \rangle = \alpha \langle f, g \rangle + \beta \langle \tilde{f}, g \rangle, & \alpha, \beta \in \mathbb{C}, \ f, \tilde{f} \in X_{\mathbb{C}}, \ g \in Y_{\mathbb{C}}, \\ \langle f, \alpha g + \beta \tilde{g} \rangle = \overline{\alpha} \langle f, g \rangle + \overline{\beta} \langle f, \tilde{g} \rangle, & \alpha, \beta \in \mathbb{C}, \ f \in X_{\mathbb{C}}, \ g, \tilde{g} \in Y_{\mathbb{C}}, \end{cases} \tag{1.3}$$

is called *a sesquilinear form* on $X_{\mathbb{C}} \times Y_{\mathbb{C}}$. Furthermore, a sesquilinear form on

$X_{\mathbb{C}} \times Y_{\mathbb{C}}$ is called *a duality product* if it satisfies

$$|\langle f, g \rangle| \le \|f\|_X \|g\|_Y, \qquad f \in X_{\mathbb{C}}, \ g \in Y_{\mathbb{C}},$$

$$\|f\|_X = \sup_{\|g\|_Y \le 1} |\langle f, g \rangle|, \qquad f \in X_{\mathbb{C}},$$

$$\|g\|_Y = \sup_{\|f\|_X \le 1} |\langle f, g \rangle|, \qquad g \in Y_{\mathbb{C}}.$$

When there exists such a duality product for two Banach spaces $X_{\mathbb{C}}$ and $Y_{\mathbb{C}}$, the space $Y_{\mathbb{C}}$ is called *an adjoint space* of $X_{\mathbb{C}}$. Obviously, this relation is symmetric and the adjointness is a property for pairs of Banach spaces; so, it is natural to say that $\{X_{\mathbb{C}}, Y_{\mathbb{C}}\}$ forms *an adjoint pair* with duality product $\langle \cdot, \cdot \rangle$. If $X_{\mathbb{C}}$ is a Hilbert space, then $X_{\mathbb{C}}$ is clearly adjoint to itself with its inner product $(\cdot, \cdot)_X$.

Let $X_{\mathbb{C}}$ be a Banach space, and $X_{\mathbb{C}}'$ be its dual space. We introduce the function $\langle f, \varphi \rangle = \varphi(f)$ for $f \in X_{\mathbb{C}}$ and $\varphi \in X_{\mathbb{C}}'$. Owing to (1.1), this function defines a sesquilinear form on $X_{\mathbb{C}} \times X_{\mathbb{C}}'$. Moreover, by virtue of the Hahn–Banach extension theorem, this function gives a duality product on $X_{\mathbb{C}} \times X_{\mathbb{C}}'$. Hence, $\{X_{\mathbb{C}}, X_{\mathbb{C}}\}$ forms an adjoint pair.

As verified above, $\{X_{\mathbb{C}}, X_{\mathbb{C}}'\}$ is an adjoint pair. So is $\{X_{\mathbb{C}}', X_{\mathbb{C}}''\}$. Therefore, $X_{\mathbb{C}}'$ has at least two adjoint spaces $X_{\mathbb{C}}$ and $X_{\mathbb{C}}''$ if $X_{\mathbb{C}}$ is a non-reflexive space. In order to denote an unspecified adjoint space of $X_{\mathbb{C}}$ we shall use the notation $X_{\mathbb{C}}^*$. Therefore, $X_{\mathbb{C}}^*$ is not uniquely determined from $X_{\mathbb{C}}$, quite differently from its dual space $X_{\mathbb{C}}'$.

Let $\{X_{\mathbb{C}}, X_{\mathbb{C}}^*\}$ be an adjoint pair with duality product $\langle \cdot, \cdot \rangle_{X \times X^*}$. For each $g \in X_{\mathbb{C}}^*$, the linear functional $\langle \cdot, g \rangle_{X \times X^*}$ is continuous on $X_{\mathbb{C}}$; therefore, a correspondence $J : g \mapsto Jg$ from $X_{\mathbb{C}}^*$ into $X_{\mathbb{C}}'$ is defined by the relation

$$[Jg](f) = \langle f, g \rangle_{X \times X^*} \qquad \text{for all } f \in X_{\mathbb{C}}. \tag{1.4}$$

Owing to (1.1), J is a linear operator on $X_{\mathbb{C}}^*$. In addition, it is easy to verify that J is an isometry from $X_{\mathbb{C}}^*$ onto $J(X_{\mathbb{C}}^*) \subset X_{\mathbb{C}}'$. Furthermore, when $X_{\mathbb{C}}$ is a reflexive Banach space, J is verified to be onto $X_{\mathbb{C}}'$.

Theorem 1.2 *If $X_{\mathbb{C}}$ is a reflexive Banach space, then the mapping $J : X_{\mathbb{C}}^* \to X_{\mathbb{C}}'$ defined by* (1.4) *is an isometric isomorphism.*

When $X_{\mathbb{C}}$ is a Hilbert space, $\{X_{\mathbb{C}}, X_{\mathbb{C}}\}$ is an adjoint pair with its inner product. So, the two definitions (1.2) and (1.4) for the operator J are consistent.

1.1.3 Interpolation of Spaces

Let $X_{\mathbb{C}}$ and $Z_{\mathbb{C}}$ be two complex Banach spaces with norm $\| \cdot \|_X$ and $\| \cdot \|_Z$ respectively. Assume that $Z_{\mathbb{C}} \subset X_{\mathbb{C}}$ densely and continuously.

Let $S = \{z; \ 0 < \operatorname{Re} z < 1\}$ be the strip in the plane $\mathbb{C}$. By $\mathcal{H}(X_\mathbb{C}, Z_\mathbb{C})$ we denote a space of analytic functions with the following properties:

$\mathcal{H}(X_\mathbb{C}, Z_\mathbb{C}) = \{F(z); \ F(z)$ is an analytic function for $z \in S$ with values in $X_\mathbb{C}$,

is a bounded and continuous function for $z \in \overline{S}$ with values in $X_\mathbb{C}$,

and is a bounded and continuous function for $z = 1 + iy$ with

values in $Z_\mathbb{C}\}$.

The space $\mathcal{H}(X_\mathbb{C}, Z_\mathbb{C})$ becomes a Banach space by the norm

$$\|F\|_\mathcal{H} = \max \left\{ \sup_{-\infty < y < \infty} \|F(iy)\|_X, \ \sup_{-\infty < y < \infty} \|F(1 + iy)\|_Z \right\}.$$

For $0 \leq \theta \leq 1$, we define the space $[X_\mathbb{C}, Z_\mathbb{C}]_\theta$ in the following way:

$$[X_\mathbb{C}, Z_\mathbb{C}]_\theta = \{u \in X_\mathbb{C}; \ \text{there exists a function } F \in \mathcal{H}(X_\mathbb{C}, Z_\mathbb{C})$$

$$\text{such that } u = F(\theta)\},$$

together with its norm

$$\|u\|_\theta = \inf_{F \in \mathcal{H}, \ F(\theta)=u} \|F\|_\mathcal{H}.$$

Then, $[X_\mathbb{C}, Z_\mathbb{C}]_\theta$ is a Banach space and is seen to possess the following basic properties:

1. $[X_\mathbb{C}, Z_\mathbb{C}]_0 = X_\mathbb{C}$ and $[X_\mathbb{C}, Z_\mathbb{C}]_1 = Z_\mathbb{C}$ with isometries.
2. For $0 < \theta < 1$, $Z_\mathbb{C} \subset [X_\mathbb{C}, Z_\mathbb{C}]_\theta \subset X_\mathbb{C}$ densely and continuously.
3. For $0 < \theta < 1$, $\|u\|_\theta \leq \|u\|_X^{1-\theta} \|u\|_Z^\theta$ for all $u \in Z_\mathbb{C}$.
4. For $0 < \theta < \theta' < 1$, $[X_\mathbb{C}, Z_\mathbb{C}]_{\theta'} \subset [X_\mathbb{C}, Z_\mathbb{C}]_\theta$ continuously.

1.1.4 Triplets of Spaces

Let $X_\mathbb{C}$ be a complex Hilbert space with inner product $(\cdot, \cdot)_X$ and norm $\|\cdot\|_X$ and let $Z_\mathbb{C}$ be a second complex Hilbert space with norm $\|\cdot\|_Z$. Assume that $Z_\mathbb{C} \subset X_\mathbb{C}$ densely and continuously.

By the techniques of extrapolation of spaces, we can construct a third space $Z_{\mathbb{C}}^*$ which is a complex Banach space with norm $\|\cdot\|_{Z^*}$ and enjoys the following properties:

1. $Z_{\mathbb{C}} \subset X_{\mathbb{C}} \subset Z_{\mathbb{C}}^*$ with dense and continuous embeddings.
2. $\{Z_{\mathbb{C}}, Z_{\mathbb{C}}^*\}$ forms an adjoint pair with duality product $\langle \cdot, \cdot \rangle_{Z \times Z^*}$, i.e.,

$$|\langle u, \varphi \rangle_{Z \times Z^*}| \leq \|u\|_Z \|\varphi\|_{Z^*}, \qquad u \in Z_{\mathbb{C}}, \ \varphi \in Z_{\mathbb{C}}^*, \tag{1.5}$$

$$\|u\|_Z = \sup_{\|\varphi\|_{Z^*} \leq 1} |\langle u, \varphi \rangle_{Z \times Z^*}|, \qquad u \in Z_{\mathbb{C}}, \tag{1.6}$$

$$\|\varphi\|_{Z^*} = \sup_{\|u\|_Z \leq 1} |\langle u, \varphi \rangle_{Z \times Z^*}|, \qquad \varphi \in Z_{\mathbb{C}}^*. \tag{1.7}$$

3. The duality product satisfies

$$\langle u, f \rangle_{Z \times Z^*} = (u, f)_X, \qquad u \in Z_{\mathbb{C}}, \ f \in X_{\mathbb{C}}. \tag{1.8}$$

Such a space $Z_{\mathbb{C}}^*$ is, as a matter of fact, uniquely determined from $Z_{\mathbb{C}} \subset X_{\mathbb{C}}$. We call the spaces $Z_{\mathbb{C}} \subset X_{\mathbb{C}} \subset Z_{\mathbb{C}}^*$ *a triplet of complex spaces*.

The property (1) naturally yields inclusions for their dual spaces such that $(Z_{\mathbb{C}}^*)' \subset X_{\mathbb{C}}' \subset Z_{\mathbb{C}}'$ densely and continuously. Let J be the isometric isomorphism defined by (1.2) or (1.4). Then, J is an isometric isomorphism not only from $X_{\mathbb{C}}$ onto $X_{\mathbb{C}}'$ and from $Z_{\mathbb{C}}^*$ onto $Z_{\mathbb{C}}'$ but also from $Z_{\mathbb{C}}$ onto $(Z_{\mathbb{C}}^*)'$ and it holds that

$$\begin{cases} [J\varphi](u) = \langle u, \varphi \rangle_{Z \times Z^*} & \text{for any} \quad u \in Z_{\mathbb{C}}, \ \varphi \in Z_{\mathbb{C}}^*, \\ [Jg](f) = (f, g)_X & \text{for any} \quad f \in X_{\mathbb{C}}, \ g \in X_{\mathbb{C}}, \\ [Ju](\varphi) = \langle \varphi, u \rangle_{Z^* \times Z} & \text{for any} \quad u \in Z_{\mathbb{C}}, \ \varphi \in Z_{\mathbb{C}}^*. \end{cases}$$

The space $X_{\mathbb{C}}$ is always characterized as the interpolation space

$$[Z_{\mathbb{C}}^*, Z_{\mathbb{C}}]_{\frac{1}{2}} = X_{\mathbb{C}} \quad \text{with norm equivalence.} \tag{1.9}$$

1.1.5 Sectorial Operators

Let $X_{\mathbb{C}}$ be a complex Banach space with norm $\|\cdot\|_X$. A densely defined, closed linear operator A of $X_{\mathbb{C}}$ is called *a sectorial operator* if its spectrum $\sigma(A)$ is contained in an open sectorial domain

$$\sigma(A) \subset \Sigma_\omega = \{\lambda \in \mathbb{C}; \ |\arg \lambda| < \omega\}, \qquad 0 < \omega \leq \pi, \tag{1.10}$$

and if its resolvent satisfies the estimate

$$\|(\lambda - A)^{-1}\| \leq \frac{M}{|\lambda|} \qquad \text{for } \lambda \notin \Sigma_\omega, \tag{1.11}$$

with some constant $M \geq 1$.

Since (1.10) implicitly means that $0 \notin \sigma(A)$, i.e., $0 \in \rho(A)$, the condition implies that A has a bounded inverse $A^{-1} \in \mathcal{L}(X_{\mathbb{C}})$. Under (1.10)–(1.11), if $\omega'(< \omega)$ is sufficiently close to ω, then there exists a constant $M' > M$ such that (1.10)–(1.11) but ω' and M' substitute for ω and M are valid. So, the infimum of the angles ω for which (1.10)–(1.11) hold true is called the angle of A and is denoted by ω_A. Then, we have $0 \leq \omega_A < \pi$ and $\sigma(A) \subset \overline{\Sigma_{\omega_A}}$.

(I) *Fractional Powers* Let A be a sectorial operator of $X_{\mathbb{C}}$ with angle ω_A and let $\omega_A < \omega < \pi$. For $\text{Re } z > 0$, its fractional power of exponent $-z$ is defined by the integral

$$A^{-z} = \frac{1}{2\pi i} \int_\Gamma \lambda^{-z} (\lambda - A)^{-1} d\lambda \tag{1.12}$$

in the space $\mathcal{L}(X_{\mathbb{C}})$ with a suitable integral contour Γ lying in $\rho(A)$.

Analyticity of A^{-z} for $\text{Re } z > 0$ and invertibility of A^{-1} implies that the bounded linear operator A^{-z} is also invertible for every $\text{Re } z > 0$; therefore, $A^z = [A^{-z}]^{-1}$ is defined as a closed linear operator of $X_{\mathbb{C}}$.

We can list the basic properties of the fractional powers A^x $(-\infty < x < \infty)$:

1. For $0 < x < \infty$, A^{-x} is an analytic function with values in $\mathcal{L}(X_{\mathbb{C}})$.
2. It holds that $A^{-x} A^{-x'} = A^{-(x+x')}$ for any $0 < x, x' < \infty$.
3. For $0 < x < \infty$, A^x is a densely defined, closed linear operator of $X_{\mathbb{C}}$.
4. It holds that $A^x A^{x'} = A^{x+x'}$ for any $0 < x, x' < \infty$ in the sense of product for unbounded linear operators.
5. For each $f \in X_{\mathbb{C}}$, as $x \searrow 0$, $A^{-x} f$ tends to f in $X_{\mathbb{C}}$.

In view of the property (5), we are led to define $A^0 = I$.

Let $0 < \theta < 1$. It is possible to show that A^θ is also a sectorial operator of X with angle $\omega_{A^\theta} \leq \theta \omega_A$. So, the fractional powers of A^θ are considered; indeed, we can verify that

$$[A^\theta]^{-x} = A^{-\theta x} \quad \text{and} \quad [A^\theta]^x = A^{\theta x} \quad \text{for } 0 < x < \infty. \tag{1.13}$$

Finally, we want to notice a reiteration property which we need in Chap. 4.

Theorem 1.3 *Let A be a positive definite self-adjoint operator of a Hilbert space $X_{\mathbb{C}}$. Let $0 \leq \theta_0 < \theta_1 \leq 1$. Then, for $\theta_0 < \theta < \theta_1$, it holds that*

$$\mathcal{D}(A^\theta) = [\mathcal{D}(A^{\theta_0}), \mathcal{D}(A^{\theta_1})]_{\theta'}, \quad \text{where } (1 - \theta')\theta_0 + \theta'\theta_1 = \theta. \tag{1.14}$$

This property has already been proved by [LM68, Théorème I.6.1]. However, some straightforward proof of utilizing the law (1.13) is presented in the proof of [Yag, Theorem 1.3].

(II) *Analytic Semigroups* Let A be a sectorial operator of $X_{\mathbb{C}}$ with angle $\omega_A < \frac{\pi}{2}$. Then, the analytic semigroup generated by $-A$ is defined by the integral

$$e^{-tA} = \frac{1}{2\pi i} \int_{\Gamma} e^{-t\lambda}(\lambda - A)^{-1} d\lambda, \qquad 0 < t < \infty, \qquad (1.15)$$

in $\mathcal{L}(X_{\mathbb{C}})$ with an integral contour $\Gamma : \lambda = re^{\pm i\omega}$ $(\omega_A < \omega < \frac{\pi}{2})$.

The semigroup e^{-tA} has the following basic properties:

1. For $0 < t < \infty$, e^{-tA} is an analytic function with values in $\mathcal{L}(X_{\mathbb{C}})$.
2. It holds that $e^{-tA}e^{-t'A} = e^{-(t+t')A}$ for any $0 < t,\, t' < \infty$.
3. For $0 < t < \infty$, $\mathcal{R}(e^{-tA}) \subset \mathcal{D}(A)$ and the operator $Ae^{-tA} = e^{-tA}A$ is bounded on $X_{\mathbb{C}}$.
4. The norm estimates $\|e^{-tA}\|_{\mathcal{L}(X)} \leq Ce^{-\delta t}$ and $\|Ae^{-tA}\|_{\mathcal{L}(X)} \leq C(1 + t^{-1})e^{-\delta t}$ hold for $0 < t < \infty$ with some $\delta > 0$ and some $C \geq 1$.
5. It holds that $\frac{d}{dt}e^{-tA} = -Ae^{-tA}$ for $0 < t < \infty$.
6. For each $f \in X_{\mathbb{C}}$, as $t \searrow 0$, $e^{-tA}f$ tends to f in $X_{\mathbb{C}}$.

In view of the property (6), we naturally define that $e^{-0A} = I$.

1.1.6 Abstract Parabolic Evolution Equations

Let $X_{\mathbb{C}}$ be a complex Banach space with norm $\|\cdot\|_X$. Let A be a sectorial operator of $X_{\mathbb{C}}$ with angle $\omega_A < \frac{\pi}{2}$. Consider the Cauchy problem for a semilinear evolution equation

$$\begin{cases} u' + Au = f(u), & 0 < t < \infty, \\ u(0) = u_0, \end{cases} \qquad (1.16)$$

in the space $X_{\mathbb{C}}$. Here, $f : \mathcal{D}(f) \to X_{\mathbb{C}}$ is a given nonlinear operator, u_0 is an initial value in $X_{\mathbb{C}}$, and $u = u(t)$ is an unknown function with values in $X_{\mathbb{C}}$.

For the operator f, assume that $\mathcal{D}(A^\eta) \subset \mathcal{D}(f)$ with some exponent $0 \leq \eta < 1$ and assume that f satisfies the following Lipschitz condition

$$\|f(u) - f(v)\|_X \leq \psi(\|A^\beta u\|_X + \|A^\beta v\|_X)[\|A^\eta(u - v)\|_X$$

$$+ (\|A^\eta u\|_X + \|A^\eta v\|_X)\|A^\beta(u - v)\|_X], \qquad u, v \in \mathcal{D}(A^\eta), \qquad (1.17)$$

with respect to the fractional powers A^η and A^β, where β is a second exponent such that $0 \le \beta \le \eta < 1$ and $\psi(\cdot) > 0$ denotes some increasing function.

Then, we have existence and uniqueness for the strict solution to (1.16).

Theorem 1.4 *Under* (1.17), *for any* $u_0 \in \mathcal{D}(A^\beta)$, *there exists* $T_{u_0} > 0$ *such that* (1.16) *possesses a unique local solution on* $[0, T_{u_0}]$ *in the function space*

$$u \in \mathcal{C}((0, T_{u_0}]; \mathcal{D}(A)) \cap \mathcal{C}([0, T_{u_0}]; \mathcal{D}(A^\beta)) \cap \mathcal{C}^1((0, T_{u_0}]; X_{\mathbb{C}}) \qquad (1.18)$$

and that the local solution satisfies the estimate

$$t^{1-\beta} \|Au(t)\|_X + \|A^\beta u(t)\|_X \le C_{u_0}, \qquad 0 < t \le T_{u_0}. \qquad (1.19)$$

Here, the constant C_{u_0} *and the time* T_{u_0} *are determined depending only on the magnitude of* $\|A^\beta u_0\|_X$.

Here and in what follows, we denote by $\mathcal{C}(I; X_{\mathbb{C}})$ (resp. $\mathcal{C}^m(I; X_{\mathbb{C}})$, where $m = 1, 2, 3, \ldots$) the space of all continuous (resp. m-times continuously differentiable) functions on an interval I with values in a Banach space $X_{\mathbb{C}}$. Meanwhile, we denote by $\mathcal{B}(I; X_{\mathbb{C}})$ the space of all uniformly bounded functions on I.

1.2 Real Banach Spaces and Hilbert Spaces

This section is devoted to reviewing the notion of conjugation on complex Banach spaces and that of real subspaces of complex Banach spaces.

1.2.1 Conjugated Spaces

Let $X_{\mathbb{C}}$ be a complex Banach space with norm $\| \cdot \|_X$. Assume that $X_{\mathbb{C}}$ is equipped with a correspondence $f \mapsto \overline{f}$ from $X_{\mathbb{C}}$ into itself having the following properties:

$$\overline{f + g} = \overline{f} + \overline{g}, \qquad f, g \in X_{\mathbb{C}}, \qquad (1.20)$$

$$\overline{\alpha f} = \overline{\alpha}\, \overline{f}, \qquad \alpha \in \mathbb{C}, f \in X_{\mathbb{C}}, \qquad (1.21)$$

$$\overline{\overline{f}} = f, \qquad f \in X_{\mathbb{C}}, \qquad (1.22)$$

$$\|f\|_X = \|\overline{f}\|_X, \qquad f \in X_{\mathbb{C}}. \qquad (1.23)$$

It is easy to see that this correspondence is continuous on $X_{\mathbb{C}}$ and bijective. In particular, $\overline{0} = 0$. The vector $\overline{f}$ is called the conjugate vector of f and such a correspondence is called *a conjugation on* $X_{\mathbb{C}}$ and $X_{\mathbb{C}}$ is said to be *a conjugated space*.

For each $f \in X_{\mathbb{C}}$, we put $\operatorname{Re} f = \frac{f+\overline{f}}{2}$ and $\operatorname{Im} f = \frac{f-\overline{f}}{2i}$. Noting that $f = \operatorname{Re} f + i \operatorname{Im} f$, we call $\operatorname{Re} f$ (resp. $\operatorname{Im} f$) the real part (resp. imaginary part) of f with respect to the conjugation $f \mapsto \overline{f}$. Then, the vectors satisfying $\operatorname{Im} f = 0$ are called real vectors. By (1.20), (1.21) and (1.22), both $\operatorname{Re} f$ and $\operatorname{Im} f$ are real vectors. As noticed, 0 is also a real vector. On the other hand, the vectors satisfying $\operatorname{Re} f = 0$ are called purely imaginary vectors. Using $\operatorname{Re} f$ and $\operatorname{Im} f$, the conjugate vector $\overline{f}$ of f is written as $\overline{f} = \operatorname{Re} f - i \operatorname{Im} f$.

We want to consider the subset

$$X = \{f \in X_{\mathbb{C}}; \ \operatorname{Im} f = 0, \text{i.e., } \overline{f} = f\}. \tag{1.24}$$

By (1.20) and (1.21), X is a real linear space equipped with the norm $\| \cdot \|_X$. Therefore, X is a real Banach space, which is called *the real subspace of $X_{\mathbb{C}}$*.

As each vector $f \in X_{\mathbb{C}}$ is uniquely expressed in the form $f = \operatorname{Re} f + i \operatorname{Im} f$, where $\operatorname{Re} f, \operatorname{Im} f \in X$, $X_{\mathbb{C}}$ is algebraically decomposed into the direct sum $X + iX$. In the meantime, by (1.23) it is easily seen that the correspondence $f \leftrightarrow (\operatorname{Re} f, \operatorname{Im} f)$ is bicontinuous from $X_{\mathbb{C}}$ onto $X \times X$, together with

$$\max\{\|\operatorname{Re} f\|_X, \ \|\operatorname{Im} f\|_X\} \leq \|f\|_X \leq \|\operatorname{Re} f\|_X + \|\operatorname{Im} f\|_X, \qquad f \in X_{\mathbb{C}}.$$

In this sense, we can say that $X_{\mathbb{C}}$ is also topologically decomposed into $X + iX$.

Theorem 1.5 *If $X_{\mathbb{C}}$ is a Banach space with conjugation, then the subset X given by (1.24) is a real Banach space with the norm $\| \cdot \|_X$ and $X_{\mathbb{C}}$ is algebraically and topologically decomposed into the form $X_{\mathbb{C}} = X + iX$.*

Any conjugation of $X_{\mathbb{C}}$ naturally induces a conjugation on its dual space $X'_{\mathbb{C}}$.

Theorem 1.6 *If a Banach space $X_{\mathbb{C}}$ is equipped with a conjugation, then so is its dual space $X'_{\mathbb{C}}$.*

Proof Indeed, consider the correspondence $\varphi \mapsto \overline{\varphi}$ in $X'_{\mathbb{C}}$ such that $\overline{\varphi}(f) = \overline{\varphi(\overline{f})}$ or $\overline{\varphi}(\overline{f}) = \overline{\varphi(f)}$ for $f \in X_{\mathbb{C}}$. Then, it is easy to see that $\overline{\varphi} \in X'_{\mathbb{C}}$ and this correspondence fulfils all of (1.20)–(1.23).   □

The real subspace of $X'_{\mathbb{C}}$ is denoted by X'. Therefore, we have $X'_{\mathbb{C}} = X' + iX'$.

Now let $X_{\mathbb{C}}$ be a complex Hilbert space with inner product $(\cdot, \cdot)_X$ and where $X_{\mathbb{C}}$ is equipped with a conjugation $f \mapsto \overline{f}$. As is well known, the inner product can be represented by a sum of norms as

$$4(f, g)_X = \|f + g\|_X^2 - \|f - g\|_X^2 + i\|f + ig\|_X^2 - i\|f - ig\|_X^2, \qquad f, g \in X_{\mathbb{C}}.$$

It is then verified that the equality

$$\overline{(f, g)}_X = (\overline{f}, \overline{g})_X, \qquad f, g \in X_{\mathbb{C}}, \tag{1.25}$$

holds true, which means that the inner product $(f, g)_X$ is real for f, $g \in X$, X being the real subspace defined by (1.24). Thus, $(\cdot, \cdot)_X$ provides a real inner product of X.

Theorem 1.7 *If $X_{\mathbb{C}}$ is a Hilbert space with conjugation, then its real subspace X given by (1.24) possesses a structure of the real Hilbert space.*

Conversely, let $X_{\mathbb{R}}$ be a real Hilbert space with inner product $(\cdot, \cdot)_X$. We can then set a complex linear space $X_{\mathbb{C}} = X_{\mathbb{R}} + i X_{\mathbb{R}}$ in the usual manner. Moreover, for the pair of vectors of $X_{\mathbb{C}}$, define their inner product by means of

$$(f + ig, f' + ig')_X = (f, f')_X + i(g, f')_X - i(f, g')_X + (g, g')_X. \qquad (1.26)$$

Then, it is not difficult to verify that this manner defines a complex inner product on $X_{\mathbb{C}}$. In addition, it holds for the norm of $X_{\mathbb{C}}$ that

$$\|f + ig\|_X^2 = (f + ig, f + ig)_X = \|f\|_X^2 + \|g\|_X^2, \qquad f + ig \in X_{\mathbb{C}}. \qquad (1.27)$$

We can therefore state the converse assertion of Theorem 1.7.

Theorem 1.8 *Let a real Hilbert space $X_{\mathbb{R}}$ be given. Then, the linear space $X_{\mathbb{C}} = X_{\mathbb{R}} + i X_{\mathbb{R}}$ equipped with the inner product (1.26) is a complex Hilbert space and the mapping $f + ig \mapsto f - ig$ for f, $g \in X_{\mathbb{R}}$ defines a conjugation on $X_{\mathbb{C}}$ whose real subspace is nothing more than $X_{\mathbb{R}}$.*

Remark 1.1 However, the situation is quite different if the given space $X_{\mathbb{R}}$ is a real Banach space. If we set the quantity $\|f + ig\|_X$ for $f + ig \in X_{\mathbb{C}}$ by the same formula as (1.27), i.e., $\|f + ig\|_X^2 = \|f\|_X^2 + \|g\|_X^2$, then $\| \cdot \|_X$ defines only a quasi-norm ([Yos80, Definition 2, P. 31]) and $X_{\mathbb{C}}$ is a quasi-normed space. Therefore, by (1.27), one cannot prove the converse of Theorem 1.5.

1.2.2 Interpolation in Conjugated Spaces

Let $X_{\mathbb{C}}$ and $Z_{\mathbb{C}}$ be two complex Banach spaces with norm $\| \cdot \|_X$ and $\| \cdot \|_Z$ respectively such that $Z_{\mathbb{C}} \subset X_{\mathbb{C}}$ densely and continuously. Assume that both $X_{\mathbb{C}}$ and $Z_{\mathbb{C}}$ are equipped with conjugation. In addition, we assume that these conjugations are consistent in the sense that the conjugation $f \mapsto \overline{f}$ on $X_{\mathbb{C}}$ enjoys the properties:

$$u \in Z_{\mathbb{C}} \quad \text{if and only if} \quad \overline{u} \in Z_{\mathbb{C}}, \qquad (1.28)$$

$$\|u\|_Z = \|\overline{u}\|_Z, \qquad u \in Z_{\mathbb{C}}. \qquad (1.29)$$

For $0 \leq \theta \leq 1$, let $[X_{\mathbb{C}}, Z_{\mathbb{C}}]_\theta$ be the interpolation spaces mentioned in Subsection 1.1.3. Then, the consistent conjugation on $X_{\mathbb{C}}$ and on $Z_{\mathbb{C}}$ runs to those spaces.

Proposition 1.1 *For any $0 \leq \theta \leq 1$, it holds true that*

$$u \in [X_{\mathbb{C}}, Z_{\mathbb{C}}]_\theta \quad \text{if and only if} \quad \overline{u} \in [X_{\mathbb{C}}, Z_{\mathbb{C}}]_\theta,$$

$$\|u\|_\theta = \|\overline{u}\|_\theta, \qquad u \in [X_{\mathbb{C}}, Z_{\mathbb{C}}]_\theta.$$

For the proof, see the proof of [Yag, Proposition 1.1].

This proposition then shows that the conjugation on $X_{\mathbb{C}}$ satisfying (1.28)–(1.29) always induces a conjugation on the interpolation space $[X_{\mathbb{C}}, Z_{\mathbb{C}}]_\theta$, too, which is consistent with the original one. Then, consider the real subspace $([X_{\mathbb{C}}, Z_{\mathbb{C}}]_\theta)_{\mathbb{R}}$ of $[X_{\mathbb{C}}, Z_{\mathbb{C}}]_\theta$ defined by (1.24). Theorem 1.5 yields the following theorem.

Theorem 1.9 *The subspace $([X_{\mathbb{C}}, Z_{\mathbb{C}}]_\theta)_{\mathbb{R}}$ is a real Banach space with the norm $\|\cdot\|_\theta$ and $[X_{\mathbb{C}}, Z_{\mathbb{C}}]_\theta$ is algebraically and topologically decomposed into*

$$[X_{\mathbb{C}}, Z_{\mathbb{C}}]_\theta = ([X_{\mathbb{C}}, Z_{\mathbb{C}}]_\theta)_{\mathbb{R}} + i\,([X_{\mathbb{C}}, Z_{\mathbb{C}}]_\theta)_{\mathbb{R}}, \qquad 0 < \theta < 1. \tag{1.30}$$

In this way, we are led to introduce a definition of interpolation for real spaces.

Definition 1.1 Let two complex Banach spaces $Z_{\mathbb{C}} \subset X_{\mathbb{C}}$ densely and continuously be equipped with a consistent conjugation. Let X and Z be the real subspaces of $X_{\mathbb{C}}$ and $Z_{\mathbb{C}}$ respectively. Then, their interpolation space $[X, Z]_\theta$ is defined by

$$[X, Z]_\theta = ([X_{\mathbb{C}}, Z_{\mathbb{C}}]_\theta)_{\mathbb{R}}, \qquad 0 \leq \theta \leq 1. \tag{1.31}$$

1.2.3 Triplets of Conjugated Spaces

Let $X_{\mathbb{C}}$ be a complex Hilbert space with inner product $(\cdot, \cdot)_X$ and norm $\|\cdot\|_X$ and let $Z_{\mathbb{C}}$ be a second complex Hilbert space with norm $\|\cdot\|_Z$. Assume that $Z_{\mathbb{C}} \subset X_{\mathbb{C}}$ densely and continuously. Then, as reviewed in Subsection 1.1.4, we have a triplet $Z_{\mathbb{C}} \subset X_{\mathbb{C}} \subset Z_{\mathbb{C}}^*$ with the properties (1)–(3). In addition, assume that $X_{\mathbb{C}}$ and $Z_{\mathbb{C}}$ are equipped with a consistent conjugation in the sense of (1.28)–(1.29).

Then, we obtain the following results. For the detailed arguments, see [Yag, Subsection 1.2.3].

First, we see that the conjugation runs to the third space $Z_{\mathbb{C}}^*$. Indeed, on account of (1.7), (1.8), (1.25) and (1.29), it holds for $f \in X_{\mathbb{C}}$ that

$$\|f\|_{Z^*} = \sup_{\|u\|_Z \leq 1} |(u, f)_X| = \sup_{\|u\|_Z \leq 1} |\overline{(u, f)_X}|$$

$$= \sup_{\|u\|_Z \leq 1} |(\overline{u}, \overline{f})_X| = \sup_{\|\overline{u}\|_Z \leq 1} |(\overline{u}, \overline{f})_X| = \|\overline{f}\|_{Z^*}.$$

This means that $f \mapsto \overline{f}$ is continuous also with respect to the norm $\| \cdot \|_{Z^*}$ on $X_{\mathbb{C}}$. Since $X_{\mathbb{C}}$ is dense in $Z_{\mathbb{C}}^*$, it is possible to extend the conjugation $f \mapsto \overline{f}$ continuously over $Z_{\mathbb{C}}^*$ with the equality $\|\varphi\|_{Z^*} = \|\overline{\varphi}\|_{Z^*}$ for all $\varphi \in Z_{\mathbb{C}}^*$. Similarly, it follows from (1.8) and (1.25) that

$$\overline{\langle u, \varphi \rangle}_{Z \times Z^*} = \langle \overline{u}, \overline{\varphi} \rangle_{Z \times Z^*}, \qquad u \in Z_{\mathbb{C}}, \ \varphi \in Z_{\mathbb{C}}^*. \tag{1.32}$$

We denote the real subspaces of $Z_{\mathbb{C}}$, $X_{\mathbb{C}}$ and $Z_{\mathbb{C}}^*$ by Z, X and Z^*, respectively. It is then clear that $Z \subset X \subset Z^*$ densely and continuously. The property (1.32) means that the duality product $\langle \cdot, \cdot \rangle_{Z \times Z^*}$ is real on $Z \times Z^*$.

Second, by virtue of Theorem 1.2 we see that

$$\|u\|_Z = \sup_{\|\varphi\|_{Z^*} \le 1, \ \varphi \in Z_{\mathbb{R}}^*} |\langle u, \varphi \rangle_{Z \times Z^*}| \qquad \text{for } u \in Z, \tag{1.33}$$

$$\|\varphi\|_{Z^*} = \sup_{\|u\| \le 1, \ u \in Z_{\mathbb{R}}} |\langle u, \varphi \rangle_{Z \times Z^*}| \qquad \text{for } \varphi \in Z^*. \tag{1.34}$$

Third, let Z', X' and $(Z^*)'$ be the real subspaces of the dual spaces $Z_{\mathbb{C}}'$, $X_{\mathbb{C}}'$ and $(Z_{\mathbb{C}}^*)'$, respectively, obtained by Theorem 1.6. Then, we see that

$$\begin{cases} \text{the isometric isomorphism } J : Z_{\mathbb{C}}^* \to Z_{\mathbb{C}}' \text{ maps } Z^* \text{ onto } Z', \\[2mm] \text{the isometric isomorphism } J : X_{\mathbb{C}} \to X_{\mathbb{C}}' \text{ maps } X \text{ onto } X', \\[2mm] \text{the isometric isomorphism } J : Z_{\mathbb{C}} \to (Z_{\mathbb{C}}^*)' \text{ maps } Z \text{ onto } (Z^*)'. \end{cases} \tag{1.35}$$

Finally, by the definition (1.31), the fact (1.9) yields the interpolation property $[Z^*, Z]_{\frac{1}{2}} = X$.

As a result, we have the following theorem.

Theorem 1.10 *Let $Z_{\mathbb{C}} \subset X_{\mathbb{C}} \subset Z_{\mathbb{C}}^*$ be a triplet of complex spaces. Assume that $X_{\mathbb{C}}$ and $Z_{\mathbb{C}}$ are equipped with a consistent conjugation. Then, the conjugation naturally runs to $Z_{\mathbb{C}}^*$ which is still consistent with the original one. Furthermore, the real subspaces Z, X and Z^* of $Z_{\mathbb{C}}$, $X_{\mathbb{C}}$ and $Z_{\mathbb{C}}^*$, respectively, enjoy the following properties:*

1. *$Z \subset X \subset Z^*$ with dense and continuous embeddings.*
2. *$\{Z, Z^*\}$ forms an adjoint pair with a real duality product $\langle \cdot, \cdot \rangle_{Z \times Z^*}$ satisfying (1.33) and (1.34) instead of (1.6) and (1.7) respectively.*
3. *The isometric isomorphism J has the properties in (1.35).*
4. *The property $[Z^*, Z]_{\frac{1}{2}} = X$ holds true.*

Definition 1.2 Let $Z_{\mathbb{C}} \subset X_{\mathbb{C}} \subset Z_{\mathbb{C}}^*$ be a triplet of complex spaces with consistent conjugation as above. Let Z, X and Z^* be real subspaces of $Z_{\mathbb{C}}$, $X_{\mathbb{C}}$ and $Z_{\mathbb{C}}^*$, respectively. Then, we will call $Z \subset X \subset Z^*$ having the properties announced in Theorem 1.10 *a triplet of real spaces*.

Remark 1.2 Now, let a real Hilbert space $X_\mathbb{R}$ be given with inner product $(\cdot, \cdot)_X$ and norm $\| \cdot \|_X$ and let a second real Hilbert space $Z_\mathbb{R}$ be given with norm $\| \cdot \|_Z$. Assume that $Z_\mathbb{R} \subset X_\mathbb{R}$ densely and continuously. As shown by Theorem 1.8, there exists their complexification to complex Hilbert spaces $X_\mathbb{C} = X_\mathbb{R} + i X_\mathbb{R}$ and $Z_\mathbb{C} = Z_\mathbb{R} + i Z_\mathbb{R}$ respectively. It is clear that $Z_\mathbb{C} \subset X_\mathbb{C}$ densely and continuously; consequently, we have a triplet $Z_\mathbb{C} \subset X_\mathbb{C} \subset Z_\mathbb{C}^*$. Furthermore, it is also clear that $f + ig \mapsto f - ig$ for $f + ig \in X_\mathbb{C}$ defines a consistent conjugation on $X_\mathbb{C}$ and on $Z_\mathbb{C}$. Therefore, all the results obtained by Theorem 1.10 are available. In other words, if two real Hilbert spaces $X_\mathbb{R}$ and $Z_\mathbb{R}$ are given as above, then we can always construct a triplet $Z \subset X \subset Z^*$ of real spaces, where $X = X_\mathbb{R}$ and $Z = Z_\mathbb{R}$.

1.2.4 Real Sobolev–Lebesgue Spaces

Let Ω be an n-dimensional bounded domain with Lipschitz boundary.

For $1 \leq p \leq \infty$, let $L_p(\Omega; \mathbb{C})$ denote the Banach space consisting of all complex L_p-functions in Ω equipped with the usual L_p-norm. Clearly, the correspondence $f(x) \mapsto \overline{f}(x) \equiv \overline{f(x)}$ gives a conjugation on $L_p(\Omega; \mathbb{C})$ (thanks to the conjugation of $\mathbb{C}$ in spite of Remark 1.1). Then, we see that

$$L_p(\Omega; \mathbb{C}) = L_p(\Omega; \mathbb{R}) + i L_p(\Omega; \mathbb{R})$$

in the sense of Theorem 1.5, where $L_p(\Omega; \mathbb{R})$ is the space of real L_p-functions in Ω.

For $1 < p < \infty$ and integral $m = 0, 1, 2, \ldots$, the complex Sobolev space $H_p^m(\Omega; C)$ of order m is characterized by

$$H_p^m(\Omega; \mathbb{C}) = \{u \in L_p(\Omega; \mathbb{C}); \ D^\alpha u \in L_p(\Omega; \mathbb{C}) \ \text{for all } \alpha \text{ up to } |\alpha| \leq m\},$$

here $\alpha = (\alpha_1, \alpha_2, \ldots, \alpha_n)$ is a multi-index, $|\alpha|$ stands for $\sum_{i=1}^{n} \alpha_i$, and $D^\alpha u = D_{x_1}^{\alpha_1} D_{x_2}^{\alpha_2} \cdots D_{x_n}^{\alpha_n} u$ is the derivative of u in the distribution sense. Its norm is given by $\|u\|_{H_p^m} = \sum_{0 \leq |\alpha| \leq m} \|D^\alpha u\|_{L_p}$. Clearly, the conjugation on $L_p(\Omega; \mathbb{C})$ is consistent with that on $H_p^m(\Omega; \mathbb{C})$. As for $L_p(\Omega; \mathbb{C})$, we have

$$H_p^m(\Omega; \mathbb{C}) = H_p^m(\Omega; \mathbb{R}) + i H_p^m(\Omega; \mathbb{R})$$

in the sense of Theorem 1.5, where $H_p^m(\Omega; \mathbb{R})$ is the real Sobolev space of order m.

For $1 < p < \infty$ and fractional s such that $m < s < m+1$, the complex Lebesgue space $H_p^s(\Omega; \mathbb{C})$ of order s is known to be characterized as

$$H_p^s(\Omega; \mathbb{C}) = [H_p^m(\Omega; \mathbb{C}), H_p^{m+1}(\Omega; \mathbb{C})]_\theta, \qquad \theta = s - m.$$

Then, Theorem 1.9 is available.

It follows by (1.30) that $H_p^s(\Omega; \mathbb{C}) = H_p^s(\Omega; \mathbb{R}) + i H_p^s(\Omega; \mathbb{R})$ algebraically and topologically, where $H_p^s(\Omega; \mathbb{R})$ is the real Lebesgue space of order s. In particular, we know that $u \in H_p^s(\Omega; \mathbb{C})$ if and only if $\mathrm{Re}\,u$, $\mathrm{Im}\,u \in H_p^s(\Omega; \mathbb{R})$ and it holds that

$$\max\{\|\mathrm{Re}\,u\|_{H_p^s}, \|\mathrm{Im}\,u\|_{H_p^s}\} \leq \|u\|_{H_p^s} \leq \|\mathrm{Re}\,u\|_{H_p^s} + \|\mathrm{Im}\,u\|_{H_p^s}, \quad u \in H_p^s(\Omega; \mathbb{C}).$$

Furthermore, according to Definition 1.1, we have

$$H_p^s(\Omega; \mathbb{R}) = [H_p^m(\Omega; \mathbb{R}), H_p^{m+1}(\Omega; \mathbb{R})]_\theta, \qquad \theta = s - m.$$

1.3   Real Sectorial Operators

In conjugated complex Banach spaces, we introduce a notion of real linear operators acting in the real subspaces, furthermore, that of real sectorial operators.

1.3.1   Real Operators in Conjugated Spaces

Let $X_\mathbb{C}$ be a complex Banach space with norm $\|\cdot\|_X$ and with conjugation $f \mapsto \overline{f}$, X being its real subspace. Let A denote a complex linear operator from $\mathcal{D}(A) \subset X_\mathbb{C}$ into $X_\mathbb{C}$.

Assume that an operator A satisfies the following two conditions:

$$u \in \mathcal{D}(A) \quad \text{if and only if} \quad \overline{u} \in \mathcal{D}(A); \tag{1.36}$$

$$\overline{Au} = A\overline{u} \qquad \text{for } u \in \mathcal{D}(A). \tag{1.37}$$

Following (1.24), let us define

$$[\mathcal{D}(A)]_\mathbb{R} = \{u \in \mathcal{D}(A); \ u = \overline{u}, \text{i.e., } \mathrm{Im}\,u = 0\} \subset X.$$

Then, (1.36) implies that $[\mathcal{D}(A)]_\mathbb{R}$ is a real linear subspace of X and that $\mathcal{D}(A)$ is decomposed algebraically into

$$\mathcal{D}(A) = [\mathcal{D}(A)]_\mathbb{R} + i[\mathcal{D}(A)]_\mathbb{R}.$$

In addition, (1.37) implies that, if $u \in [\mathcal{D}(A)]_\mathbb{R}$, then $\overline{Au} = A\overline{u} = Au$ i.e., $Au \in X$, which shows that A maps $[\mathcal{D}(A)]_\mathbb{R}$ into X.

Therefore, we notice the following fact.

Theorem 1.11 *Let a complex linear operator A satisfy (1.36)–(1.37). Then, $[\mathcal{D}(A)]_\mathbb{R}$ is a real linear subspace of X such that $\mathcal{D}(A) = [\mathcal{D}(A)]_\mathbb{R} + i[\mathcal{D}(A)]_\mathbb{R}$ algebraically, and A is a real linear operator from $[\mathcal{D}(A)]_\mathbb{R}$ into X.*

So, when a complex linear operator A enjoys the properties (1.36)–(1.37), A is called *a real operator of* $X_{\mathbb{C}}$. When we consider A as a real linear operator acting in X, we will denote it by the same notation A if there is no fear of confusion.

If A is a densely defined, closed linear operator of $X_{\mathbb{C}}$ which is real, then A is, at the same time, a densely defined, closed linear operator of X with the domain $\mathcal{D}(A) \cap X = [\mathcal{D}(A)]_{\mathbb{R}}$.

1.3.2 Sectorial Operators in Conjugated Spaces

Now, consider a sectorial operator A of $X_{\mathbb{C}}$, namely, let A be a densely defined, closed linear operator of $X_{\mathbb{C}}$ satisfying (1.10)–(1.11). When A is, at the same time, a real operator of $X_{\mathbb{C}}$, we say that A is *a real sectorial operator of* $X_{\mathbb{C}}$.

The resolvent of a real sectorial operator has the following properties.

Proposition 1.2 *Let A be a real sectorial operator of* $X_{\mathbb{C}}$. *Then,*

$$\lambda \in \rho(A) \quad \text{if and only if} \quad \overline{\lambda} \in \rho(A), \tag{1.38}$$

$$\overline{(\lambda - A)^{-1} f} = (\overline{\lambda} - A)^{-1} \overline{f} \quad \text{for } \lambda \in \rho(A), \ f \in X_{\mathbb{C}}. \tag{1.39}$$

For the proof, see the proof of [Yag, Proposition 1.2].

By the properties (1.38)–(1.39) of the real sectorial operator A, we can verify that its functional calculus also enjoys the properties (1.36)–(1.37).

(I) *Fractional Powers* Let A be a real sectorial operator of $X_{\mathbb{C}}$. According to (1.12), for $0 < x < \infty$, $A^{-x} f$ is given by the integral

$$A^{-x} f = \frac{1}{2\pi i} \int_{\Gamma} \lambda^{-x} (\lambda - A)^{-1} f \, d\lambda, \qquad f \in X_{\mathbb{C}},$$

in the space $X_{\mathbb{C}}$. Taking a conjugation of this equality, we obtain by (1.39) that

$$\overline{A^{-x} f} = -\frac{1}{2\pi i} \int_{\Gamma} (\overline{\lambda})^{-x} (\overline{\lambda} - A)^{-1} \overline{f} \, d\overline{\lambda}$$

$$= -\frac{1}{2\pi i} \int_{-\Gamma} \lambda^{-x} (\lambda - A)^{-1} \overline{f} \, d\lambda = A^{-x} \overline{f}.$$

Here, we used the fact that, when the integral contour Γ is symmetric with respect to the real axis, the transform $\lambda \in \Gamma \mapsto \overline{\lambda} \in \Gamma$ yields an integration along the same contour but in the opposite orientation, i.e., along $-\Gamma$. Hence, A^{-x} satisfies (1.37) and A^{-x} is a real bounded operator of $X_{\mathbb{C}}$.

Since $u \in \mathcal{R}(A^{-x})$ if and only if $\bar{u} \in \mathcal{R}(A^{-x})$, we see that $u \in \mathcal{D}(A^x)$ if and only if $\bar{u} \in \mathcal{D}(A^x)$. In addition, it holds that $\overline{A^x u} = A^x \bar{u}$ for $u \in \mathcal{D}(A^x)$. By the definition (1.36)–(1.37), A^x is deduced to be a real operator of $X_{\mathbb{C}}$. Consequently,

A^x is a densely defined, closed linear operator of X

$$\text{with the domain } \mathcal{D}(A^x) \cap X = [\mathcal{D}(A^x)]_{\mathbb{R}}. \qquad (1.40)$$

We shall see that the fractional powers of real sectorial operators are very effective in setting up the triplets of real spaces in our applications.

In view of (1.30) and (1.40), we have a real version of (1.14).

Theorem 1.12 *Let $X_{\mathbb{C}}$ be a conjugated Hilbert space and let A be a positive definite self-adjoint operator of $X_{\mathbb{C}}$ which is real in $X_{\mathbb{C}}$. Let $0 \leq \theta_0 < \theta_1 \leq 1$. Then, for $\theta_0 < \theta < \theta_1$, it holds that*

$$[\mathcal{D}(A^\theta)]_{\mathbb{R}} = \left([\mathcal{D}(A^{\theta_0}), \mathcal{D}(A^{\theta_1})]_{\theta'}\right)_{\mathbb{R}}, \quad \text{where } (1 - \theta')\theta_0 + \theta'\theta_1 = \theta. \qquad (1.41)$$

(II) *Analytic Semigroups* Let A be a real sectorial operator of $X_{\mathbb{C}}$ whose angle is such that $\omega_A < \frac{\pi}{2}$. By the same arguments as for the fractional powers A^{-x}, the semigroup e^{-tA} which is given by the integral (1.15) is seen to satisfy the condition (1.37) and to be a real operator of $X_{\mathbb{C}}$. Consequently, e^{-tA} is a bounded linear operator acting on X and

$$e^{-tA} \text{ defines an analytic semigroup on } X. \qquad (1.42)$$

1.4 Operators Associated with Sesquilinear Forms

As is well known, to make use of sesquilinear forms in the framework of triplets of spaces is the most convenient manner for determining sectorial operators (see Dautray–Lions [DL84b, Chapitre VII]). These techniques are still very convenient for determining various real sectorial operators.

1.4.1 Real Sesquilinear Forms

Let $X_{\mathbb{C}}$ be a complex Hilbert space with inner product $(\cdot, \cdot)_X$ and norm $\| \cdot \|_X$ and let $Z_{\mathbb{C}}$ be a second complex Hilbert space with norm $\| \cdot \|_Z$, $Z_{\mathbb{C}}$ being embedded in $X_{\mathbb{C}}$ densely and continuously. Let $Z_{\mathbb{C}} \subset X_{\mathbb{C}} \subset Z_{\mathbb{C}}^*$ be the triplet of complex spaces.

Consider a sesquilinear form $a(u, v)$ defined on $Z_\mathbb{C} \times Z_\mathbb{C}$ (see (1.3)). Assume that $a(u, v)$ satisfies the conditions:

$$\begin{cases} |a(u, v)| \leq M \|u\|_Z \|v\|_Z & \text{for } u, v \in Z_\mathbb{C}, \\ \operatorname{Re} a(u, u) \geq \delta \|u\|_Z^2 & \text{for } u \in Z_\mathbb{C}, \end{cases} \tag{1.43}$$

with some constants $M > 0$ and $\delta > 0$.

For each fixed $u \in Z_\mathbb{C}$, the correspondence $v \mapsto \overline{a(u, v)}$ is a bounded linear functional. Then, by virtue of Theorem 1.2, there exists a vector $\mathcal{A}u \in Z_\mathbb{C}^*$ determined from u such that $\overline{a(u, v)} = \langle v, \mathcal{A}u \rangle_{Z \times Z^*}$ for all $v \in Z_\mathbb{C}$, i.e., $a(u, v) = \langle \mathcal{A}u, v \rangle_{Z^* \times Z}$, $\forall v \in Z_\mathbb{C}$. Under (1.43), the correspondence $u \mapsto \mathcal{A}u$ is seen to be a linear isomorphism from $Z_\mathbb{C}$ onto $Z_\mathbb{C}^*$. Furthermore, when $\mathcal{A}$ is considered as a closed linear operator acting in $Z_\mathbb{C}^*$, $\mathcal{A}$ is shown to be a sectorial operator of angle $\omega_\mathcal{A} < \frac{\pi}{2}$. In addition, its part $A = \mathcal{A}_{|X_\mathbb{C}}$ in $X_\mathbb{C}$, which is defined by the domain $\mathcal{D}(A) = \{u \in Z_\mathbb{C}; \mathcal{A}u \in X_\mathbb{C}\}$, is also shown to be a densely defined, closed linear operator and moreover to be a sectorial operator of $X_\mathbb{C}$ of angle $\omega_A < \frac{\pi}{2}$.

Now, assume that a consistent conjugation $f \mapsto \overline{f}$ is equipped on $X_\mathbb{C}$ and on $Z_\mathbb{C}$ (see (1.28)–(1.29)). As noticed in Theorem 1.10, this conjugation runs to that on $Z_\mathbb{C}^*$ which is still consistent. Let $Z \subset X \subset Z^*$ be the triplet of real spaces (see Definition 1.2).

We want to consider a sesquilinear form $a(u, v)$ on $Z_\mathbb{C} \times Z_\mathbb{C}$ satisfying

$$\overline{a(u, v)} = a(\overline{u}, \overline{v}) \qquad \text{for } u, v \in Z_\mathbb{C}. \tag{1.44}$$

This condition obviously implies that $a(u, v)$ is real if u and v are real in $Z_\mathbb{C}$, i.e., $u, v \in Z$. In this sense, a sesquilinear form satisfying (1.44) is called *a real sesquilinear form*.

By the property (1.32), it follows from (1.44) that

$$\overline{\langle \mathcal{A}u, v \rangle}_{Z^* \times Z} = \langle \overline{\mathcal{A}u}, \overline{v} \rangle_{Z^* \times Z} = \overline{a(u, \overline{v})} = a(\overline{u}, v) = \langle \mathcal{A}\overline{u}, v \rangle_{Z^* \times Z},$$

which yields the property $\overline{\mathcal{A}u} = \mathcal{A}\overline{u}$ for $u \in Z_\mathbb{C}$. The definition (1.36)–(1.37) then shows that $\mathcal{A}$ is a real operator of $Z_\mathbb{C}^*$. In particular, $\mathcal{A}$ defines a real linear operator from Z onto Z^*.

Similarly, A is also verified to be a real operator of $X_\mathbb{C}$. Indeed, $u \in \mathcal{D}(A)$ if and only if $\mathcal{A}u \in X_\mathbb{C}$; then, since $\overline{\mathcal{A}u} = \mathcal{A}\overline{u}$, this is equivalent to $\mathcal{A}\overline{u} \in X_\mathbb{C}$; hence, $u \in \mathcal{D}(A)$ if and only if $\overline{u} \in D(A)$. Hence, A defines a real linear operator from $[\mathcal{D}(A)]_\mathbb{R}$ onto X.

Theorem 1.13 *Let $Z_\mathbb{C} \subset X_\mathbb{C} \subset Z_\mathbb{C}^*$ be a triplet equipped with consistent conjugation. Consider a real sesquilinear form $a(u, v)$ on $Z_\mathbb{C} \times Z_\mathbb{C}$. Then, the associated linear operator $\mathcal{A}$ (resp. A) to the form $a(u, v)$ is a real sectorial operator of $Z_\mathbb{C}$ (resp. $X_\mathbb{C}$) of angle $\omega_\mathcal{A} < \frac{\pi}{2}$ (resp. $\omega_A < \frac{\pi}{2}$).*

In addition to (1.43) and (1.44), assume that $a(u, v)$ is Hermitian, i.e.,

$$a(v, u) = \overline{a(u, v)}, \qquad u, v \in Z_{\mathbb{C}}. \tag{1.45}$$

This implies that $\langle \mathcal{A}u, v \rangle_{Z^* \times Z} = a(u, v) = \overline{a(v, u)} = \overline{\langle \mathcal{A}v, u \rangle}_{Z^* \times Z} = \langle u, \mathcal{A}v \rangle_{Z \times Z^*}$ for $u, v \in Z_{\mathbb{C}}$, that is, $\mathcal{A}$ is symmetric. Similarly, (1.45) implies that $(Au, v)_X = (u, Av)_X$ for $u, v \in \mathcal{D}(A)$, that is, A is a self-adjoint operator of $X_{\mathbb{C}}$. In this case, it is known that the domain of the square root of $\mathcal{A}$ (resp. A) is characterized as $\mathcal{D}(\mathcal{A}^{\frac{1}{2}}) = X_{\mathbb{C}}$ (resp. $\mathcal{D}(A^{\frac{1}{2}}) = Z_{\mathbb{C}}$). Then, due to (1.40), we observe that

$$\mathcal{D}(\mathcal{A}^{\frac{1}{2}}) \cap Z^* = [D(A^{\frac{1}{2}})]_{\mathbb{R}} = X \quad \text{and} \quad \mathcal{D}(\mathcal{A}^{\frac{1}{2}}) \cap X = [D(A^{\frac{1}{2}})]_{\mathbb{R}} = Z. \tag{1.46}$$

1.4.2 Realization of Elliptic Operators

We are ready to present using Theorem 1.13 some examples of real sectorial operators acting in suitable real function spaces which realize second-order elliptic operators. For the detailed techniques, see [Yag10, Chapter 2, Section 1].

Let Ω be a bounded domain of $\mathbb{R}^n$ with Lipschitz boundary.

Let $L_2(\Omega; \mathbb{C})$ be the complex L_2-space in Ω with norm $\| \cdot \|_{L_2}$ and let $\overset{\circ}{H}^1(\Omega; \mathbb{C})$ denote the closure of the space $\mathcal{C}_0^\infty(\Omega; \mathbb{C})$ with respect to the Sobolev norm $\| \cdot \|_{H^1}$, where $\mathcal{C}_0^\infty(\Omega; \mathbb{C})$ is the space of test functions consisting of all $\mathcal{C}^\infty$-functions with compact support. Actually, $\overset{\circ}{H}^1(\Omega; \mathbb{C})$ is characterized as a closed subspace of $H^1(\Omega; \mathbb{C})$ consisting of functions satisfying the homogeneous Dirichlet conditions $u_{|\partial\Omega} = 0$ on the boundary $\partial\Omega$ of Ω.

As is well known, $\overset{\circ}{H}^1(\Omega; \mathbb{C}) \subset L_2(\Omega; \mathbb{C}) \subset H^{-1}(\Omega; \mathbb{C})$ make a triplet, where $H^{-1}(\Omega; \mathbb{C}) (\subset \mathcal{C}_0^\infty(\Omega; \mathbb{C})')$ is the adjoint space of $\overset{\circ}{H}^1(\Omega; \mathbb{C})$ with the usual duality product in the sense of distributions.

Let $f \mapsto \overline{f}$ be the complex conjugation on $L_2(\Omega; \mathbb{C})$ which obviously satisfies (1.20)–(1.23) and is consistent with the conjugation on $\overset{\circ}{H}^1(\Omega; \mathbb{C})$. Thereby, this also runs to $H^{-1}(\Omega; \mathbb{C})$. Consequently, the three complex spaces are decomposed into the direct sums of $\overset{\circ}{H}^1(\Omega; \mathbb{C}) = \overset{\circ}{H}^1(\Omega; \mathbb{R}) + i\overset{\circ}{H}^1(\Omega; \mathbb{R})$, $L_2(\Omega; \mathbb{C}) = L_2(\Omega; \mathbb{R}) + iL_2(\Omega; \mathbb{R})$ and $H^{-1}(\Omega; \mathbb{C}) = H^{-1}(\Omega; \mathbb{R}) + iH^{-1}(\Omega; \mathbb{R})$, respectively. As shown by Theorem 1.10, we have a triplet $\overset{\circ}{H}^1(\Omega; \mathbb{R}) \subset L_2(\Omega; \mathbb{R}) \subset H^{-1}(\Omega; \mathbb{R})$ of real spaces.

Consider a sesquilinear form

$$a(u, v) = \sum_{j,k=1}^{n} \int_\Omega a_{jk}(x) \frac{\partial u}{\partial x_j} \frac{\partial \overline{v}}{\partial x_k} dx, \qquad u, v \in \overset{\circ}{H}^1(\Omega; \mathbb{C}).$$

Here we assume that the functions $a_{jk}(x)$ satisfy the conditions:

$$a_{jk} = a_{kj} \in L_\infty(\Omega; \mathbb{R}) \qquad \text{for } 1 \leq j, k \leq n, \tag{1.47}$$

$$\sum_{j,k=1}^{n} a_{jk}(x)\xi_j\xi_k \geq \delta|\xi|^2 \quad \text{for a.e. } x \in \Omega \text{ and } \forall \xi = (\xi_1, \ldots, \xi_n) \in \mathbb{R}^n, \tag{1.48}$$

with some constant $\delta > 0$.

By (1.47)–(1.48), the form $a(u, v)$ is verified to satisfy (1.43). The assumption (1.47) yields also the conditions (1.44) and (1.45) for $a(u, v)$. Therefore, all the results stated above are available to this real sesquilinear form $a(u, v)$.

Indeed, let $\mathcal{A}$ be the operator associated with $a(u, v)$. Then, $\mathcal{A}$ is a real sectorial operator of $H^{-1}(\Omega; \mathbb{C})$ with domain $\mathcal{D}(\mathcal{A}) = \overset{\circ}{H}{}^1(\Omega; \mathbb{C})$ and with angle 0. As a result, $\mathcal{A}$ acts from $\overset{\circ}{H}{}^1(\Omega; \mathbb{R})$ onto $H^{-1}(\Omega; \mathbb{R})$. This operator $\mathcal{A}$ is then regarded as a realization of the differential operator $-\sum_{j,k=1}^{n} \frac{\partial}{\partial x_k}\left[a_{jk}(x)\frac{\partial}{\partial x_j}\right]$ in $H^{-1}(\Omega; \mathbb{R})$ under the homogeneous Dirichlet boundary conditions.

In the meantime, let A be the part of $\mathcal{A}$ in $L_2(\Omega; \mathbb{C})$. Then, A is a positive definite self-adjoint operator of $L_2(\Omega; \mathbb{C})$. So, A is a positive definite self-adjoint operator of $L_2(\Omega; \mathbb{R})$ with domain $\mathcal{D}(A) \subset \overset{\circ}{H}{}^1(\Omega; \mathbb{R})$. This operator A is regarded as a realization of the differential operator $-\sum_{j,k=1}^{n} \frac{\partial}{\partial x_k}\left[a_{jk}(x)\frac{\partial}{\partial x_j}\right]$ in $L_2(\Omega; \mathbb{R})$ under the homogeneous Dirichlet boundary conditions.

As seen from (1.30) and (1.40),

$$\begin{cases} \mathcal{D}(\mathcal{A}^\theta) \cap H^{-1}(\Omega; \mathbb{R}) = \left([H^{-1}(\Omega; \mathbb{C}), \overset{\circ}{H}{}^1(\Omega; \mathbb{C})]_\theta\right)_\mathbb{R}, & 0 < \theta < 1, \\ \mathcal{D}(A^\theta) \cap L_2(\Omega; \mathbb{R}) = ([L_2(\Omega; \mathbb{C}), \mathcal{D}(A)]_\theta)_\mathbb{R}, & 0 < \theta < 1. \end{cases}$$

In particular, when $\theta = \frac{1}{2}$, (1.46) yields

$$\mathcal{D}(\mathcal{A}^{\frac{1}{2}}) \cap H^{-1}(\Omega; \mathbb{R}) = L_2(\Omega; \mathbb{R}) \quad \text{and} \quad \mathcal{D}(A^{\frac{1}{2}}) \cap L_2(\Omega; \mathbb{R}) = \overset{\circ}{H}{}^1(\Omega; \mathbb{R}).$$

$$\tag{1.49}$$

These results provide us an important basis for obtaining a complete characterization of the domains $\mathcal{D}(\mathcal{A}^\theta)$ and $\mathcal{D}(A^\theta)$ in $H^{-1}(\Omega; \mathbb{R})$ and $L_2(\Omega; \mathbb{R})$, respectively.

Next, consider $H^1(\Omega; \mathbb{C}) \subset L_2(\Omega; \mathbb{C})$ and let $H^1(\Omega; \mathbb{C}) \subset L_2(\Omega; \mathbb{C}) \subset H^1(\Omega; \mathbb{C})^*$ be a triplet generated by them. Consider a sesquilinear form

$$a(u, v) = \sum_{j,k=1}^{n} \int_\Omega a_{jk}(x)\frac{\partial u}{\partial x_j}\frac{\partial \overline{v}}{\partial x_k}\,dx + \int_\Omega c(x)u\,\overline{v}dx, \qquad u, v \in H^1(\Omega; \mathbb{C}).$$

We assume that $a_{jk}(x)$ satisfy (1.47)–(1.48) and $c(x)$ satisfies the condition

$$c \in L_\infty(\Omega; \mathbb{R}) \quad \text{and} \quad \text{ess.} \inf_{x \in \Omega} c(x) > 0. \tag{1.50}$$

As before, the form $a(u, v)$ satisfies (1.43), (1.44) and (1.45) and all the results in the preceding subsection are available to $a(u, v)$.

Indeed, let $\mathcal{A}$ be the operator associated with $a(u, v)$. Then, $\mathcal{A}$ is a real sectorial operator of $H^1(\Omega; \mathbb{C})^*$ with domain $\mathcal{D}(\mathcal{A}) = H^1(\Omega; \mathbb{C})$ and with angle 0. Consequently, $\mathcal{A}$ acts from $H^1(\Omega; \mathbb{R})$ onto $H^1(\Omega; \mathbb{R})^*$. This $\mathcal{A}$ is regarded as a realization of the differential operator $-\sum_{j,k=1}^n \frac{\partial}{\partial x_k}\left[a_{jk}(x)\frac{\partial}{\partial x_j}\right] + c(x)$ in $H^1(\Omega; \mathbb{R})^*$ under the homogeneous Neumann type boundary conditions $\sum_{j,k=1}^n a_{jk}(x)v_k(x)\frac{\partial u}{\partial x_k} = 0$ on $\partial\Omega$, where $v(x) = (v_1(x), \ldots, v_n(x))$ stands for the outer normal vector at $x \in \partial\Omega$.

Meanwhile, let A be the part of $\mathcal{A}$ in $L_2(\Omega; \mathbb{C})$. Then, A is a positive definite self-adjoint operator of $L_2(\Omega; \mathbb{C})$. So, A is also a positive definite self-adjoint operator of $L_2(\Omega; \mathbb{R})$. This operator A is regarded as a realization of the differential operator $-\sum_{j,k=1}^n \frac{\partial}{\partial x_k}\left[a_{jk}(x)\frac{\partial}{\partial x_j}\right] + c(x)$ in $L_2(\Omega; \mathbb{R})$ under the homogeneous Neumann type boundary conditions.

As before, by (1.30) and (1.40),

$$\begin{cases} \mathcal{D}(\mathcal{A}^\theta) \cap H^1(\Omega; \mathbb{R})^* = \left([H^1(\Omega; \mathbb{C})^*, H^1(\Omega; \mathbb{C})]_\theta\right)_{\mathbb{R}}, & 0 < \theta < 1, \\ \mathcal{D}(A^\theta) \cap L_2(\Omega; \mathbb{R}) = ([L_2(\Omega; \mathbb{C}), \mathcal{D}(A)]_\theta)_{\mathbb{R}}, & 0 < \theta < 1. \end{cases}$$

Analogously to (1.49), when $\theta = \frac{1}{2}$, (1.46) yields

$$\mathcal{D}(\mathcal{A}^{\frac{1}{2}}) \cap H^1(\Omega; \mathbb{R})^* = L_2(\Omega; \mathbb{R}) \quad \text{and} \quad \mathcal{D}(A^{\frac{1}{2}}) \cap L_2(\Omega; \mathbb{R}) = H^1(\Omega; \mathbb{R}). \tag{1.51}$$

1.5 Differentiation of Operators

Let X and Y be two real Banach spaces with norm $\|\cdot\|_X$ and $\|\cdot\|_Y$ respectively. Consider an operator F from X into Y.

For $u \in X$, when there exists a bounded operator $B \in \mathcal{L}(X, Y)$ such that,

$$\text{as } h \to 0 \text{ in } X, \quad F(u+h) - F(u) - Bh = o(\|h\|_X), \tag{1.52}$$

$o(\|h\|_X)$ being a vector of Y such that, as $h \to 0$ in X, $o(\|h\|_X)/\|h\|_X \to 0$ in Y, F is said to be Fréchet differentiable at u. The operator B is called a derivative of F at u and is denoted by $F'(u)$.

Furthermore, when F is Fréchet differentiable at every vector of a neighborhood O of $u \in X$ and when the mapping $u \mapsto F'(u)$ is continuous from O into $\mathcal{L}(X, Y)$, F is said to be continuously Fréchet differentiable in O.

In the meantime, for $u \in X$, when a real variable Y-valued function

$$\theta \mapsto F(u + \theta h) \text{ is differentiable at } \theta = 0 \text{ for any } h \in X, \tag{1.53}$$

F is said to be Gâteaux differentiable at u. Put $B(u, h) = \left[\frac{d}{d\theta} F(u + \theta h) \right]_{|\theta=0} \in Y$. Then, the mapping $h \mapsto B(u, h)$ is called a derivative of F at u.

Remark 1.3 If $F : X \to Y$ is Fréchet differentiable at $u \in X$, then F is obviously Gâteaux differentiable at u with the derivative $B(u, h) = F'(u)h$. Of course, the converse is not true in general. Gâteaux derivatives $B(u, h)$ may be nonlinear or discontinuous with respect to $h \in X$. Even if $B(u, h)$ is linear and continuous in h, F is not necessarily Fréchet differentiable at u.

Let $\Omega \subset \mathbb{R}^n$ be a bounded domain and let $f(x, u)$ denote a real-valued function defined for $(x, u) \in \overline{\Omega} \times \mathbb{R}$. In this section, we want to discuss differentiability for the operators of form $u \mapsto f(x, u)$ from $L_q(\Omega)$ into $L_p(\Omega)$ where $1 \leq p \leq q \leq \infty$.

By $\mathcal{C}_\infty^{0,0}(\overline{\Omega} \times \mathbb{R})$, we denote the space of real-valued functions $f(x, u)$ for $(x, u) \in \overline{\Omega} \times \mathbb{R}$ which are continuous for (x, u) and are uniformly bounded in $\overline{\Omega} \times \mathbb{R}$ and by $\| f \|_{\mathcal{C}^{0,0}} = \sup_{(x,u)\in\overline{\Omega}\times\mathbb{R}} |f(x, u)|$ its norm. By $\mathcal{C}_\infty^{0,1}(\overline{\Omega} \times \mathbb{R})$, we denote the space of functions $f(x, u) \in \mathcal{C}_\infty^{0,0}(\overline{\Omega} \times \mathbb{R})$ which are continuously differentiable for u with partial derivative $f_u(x, u) \in \mathcal{C}_\infty^{0,0}(\overline{\Omega}\times\mathbb{R})$ and by $\| f \|_{\mathcal{C}^{0,1}} = \| f \|_{\mathcal{C}^{0,0}} + \| f_u \|_{\mathcal{C}^{0,0}}$ its norm. For $0 < \sigma < 1$, $\mathcal{C}_\infty^{0,1+\sigma}(\overline{\Omega} \times \mathbb{R})$ denotes the space of functions $f(x, u) \in \mathcal{C}_\infty^{0,1}(\overline{\Omega} \times \mathbb{R})$ whose partial derivative $f_u(x, u)$ is uniformly Hölder continuous for u with exponent σ, its norm being given by $\| f \|_{\mathcal{C}^{0,1+\sigma}} = \| f \|_{\mathcal{C}^{0,1}} + \sup_{(x,u),(x,v)\in\overline{\Omega}\times\mathbb{R}} |f_u(x, u) - f_v(x, v)|/|u - v|^\sigma$.

1.5.1 Fréchet Differentiation

In this subsection, we collect some results concerning Fréchet differentiability for the operators of form $u \mapsto f(x, u)$.

Theorem 1.14 *Let* $f \in \mathcal{C}_\infty^{0,1+\sigma}(\overline{\Omega} \times \mathbb{R})$ $(0 < \sigma < 1)$. *For* $1 \leq p < q < \infty$, *the operator* $u \mapsto f(x, u)$ *from* $L_q(\Omega)$ *into* $L_p(\Omega)$ *is continuously Fréchet differentiable with the derivative* $h \mapsto f_u(x, u)h$.

Proof Consider first the case where $(1 + \sigma)p \leq q$. Let $u, h \in L_q(\Omega)$. For a.e. $x \in \Omega$, we have

$$f(x, u(x) + h(x)) - f(x, u(x)) - f_u(x, u(x))h(x)$$

$$= \int_0^1 [f_u(x, u(x) + \theta h(x)) - f_u(x, u(x))]h(x)d\theta. \tag{1.54}$$

Therefore,

$$|f(x, u(x) + h(x)) - f(x, u(x)) - f_u(x, u(x))h(x)|^p \leq \|f\|_{\mathcal{C}^{0,1+\sigma}}^p |h(x)|^{(1+\sigma)p}.$$

By integration in Ω, it follows that

$$\|f(x, u + h) - f(x, u) - f_u(x, u)h\|_{L_p}^p \leq \|f\|_{\mathcal{C}^{0,1+\sigma}}^p \|h\|_{L_{(1+\sigma)p}}^{(1+\sigma)p}$$

and that

$$\|f(x, u + h) - f(x, u) - f_u(x, u)h\|_{L_p} \leq \|f\|_{\mathcal{C}^{0,1+\sigma}} \|h\|_{L_{(1+\sigma)p}}^{1+\sigma}$$

$$\leq C\|f\|_{\mathcal{C}^{0,1+\sigma}} \|h\|_{L_q}^{1+\sigma}.$$

Hence, the condition (1.52) is fulfilled.

Next, consider the case where $(1 + \sigma)p > q$. But, if we take $\sigma' < \sigma$ so that $(1+\sigma')p = q$, then it clearly holds that $\mathcal{C}_{\infty}^{0,1+\sigma}(\overline{\Omega} \times \mathbb{R}) \subset \mathcal{C}_{\infty}^{0,1+\sigma'}(\overline{\Omega} \times \mathbb{R})$. Thereby, the desired differentiability is obvious.

Finally, let us observe the continuity of the derivative. For $u, \overline{u} \in L_q(\Omega)$, we will also denote for simplicity their Fréchet derivatives by $f_u(x, u)$ and $f_u(x, \overline{u})$ respectively. Then, since $\|[f_u(x, u) - f_u(x, \overline{u})]h\|_{L_p} \leq \|f_u(x, u) - f_u(x, \overline{u})\|_{L_r} \|h\|_{L_q}$, where $\frac{1}{r} + \frac{1}{q} = \frac{1}{p}$, we see that

$$\|f_u(x, u) - f_u(x, \overline{u})\|_{\mathcal{L}(L_q, L_p)} \leq \|f_u(x, u) - f_u(x, \overline{u})\|_{L_r}.$$

Suppose here that, as $u \to \overline{u}$ in $L_q(\Omega)$, $f_u(x, u) - f_u(x, \overline{u})$ would not converge to 0 in $L_r(\Omega)$. Then, there exists a sequence u_n tending to $\overline{u}$ such that $\|f_u(x, u_n) - f_u(x, \overline{u})\|_{L_r} \geq \varepsilon$ with some constant $\varepsilon > 0$. As $|u_n - \overline{u}|$ tends to 0 in $L_q(\Omega)$, $|u_n - \overline{u}|$ converges to 0 in measure in Ω. Since any sequence which is convergent in measure has a subsequence which is pointwise convergent almost everywhere in Ω, we can extract a subsequence $u_{n'}$ such that $u_{n'}(x)$ converges to $\overline{u}(x)$ for a.e. $x \in \Omega$. Then, the dominated convergence theorem provides that $f_u(x, u_{n'})$ converges to $f_u(x, \overline{u})$ in $L_r(\Omega)$, which is a contradiction. □

Theorem 1.15 *Let $f \in \mathcal{C}^{0,1}(\overline{\Omega} \times \mathbb{R})$. The operator $u \mapsto f(x, u)$ from $\mathcal{C}(\overline{\Omega})$ into itself is continuously Fréchet differentiable with the derivative $h \mapsto f_u(x, u)h$.*

Proof For $u, h \in \mathcal{C}(\overline{\Omega})$, it is seen from (1.54) that

$$\|f(x, u+h) - f(x, u) - f_u(x, u)h\|_{\mathcal{C}} \leq \sup_{0 \leq \theta \leq 1} \|f_u(x, u+\theta h) - f_u(x, u)\|_{\mathcal{C}} \|h\|_{\mathcal{C}}.$$

Let $\|h\|_{\mathcal{C}} < 1$. Then, since $f_u(x, \upsilon)$ is uniformly continuous for υ on the interval $[-(\|u\|_{\mathcal{C}} + 1), \|u\|_{\mathcal{C}} + 1]$, it holds true that, as $h \to 0$ in $\mathcal{C}(\overline{\Omega})$, $\|f_u(x, u + \theta h) - f_u(x, u)\|_{\mathcal{C}} \to 0$, which shows that the condition (1.52) is fulfilled.

Continuity of the derivative is also verified by similar arguments. □

As an immediate consequence of Theorem 1.14, we have the following corollary.

Corollary 1.1 *Let* $f \in \mathcal{C}_\infty^{0,1+\sigma}(\overline{\Omega} \times \mathbb{R})$ $(0 < \sigma < 1)$. *For* $q > 1$, *the operator* $u \mapsto \int_\Omega f(x, u)dx$ *from* $L_q(\Omega)$ *into* $\mathbb{R}$ *is continuously Fréchet differentiable with the derivative* $h \mapsto \int_\Omega f_u(x, u)h\, dx$.

Proof By Theorem 1.14 (with $p = 1 < q$), we know that $u \mapsto f(x, u)$ is continuously Fréchet differentiable from $L_q(\Omega)$ into $L_1(\Omega)$ with derivative $h \mapsto f_u(x, u)h$. Meanwhile, $v \mapsto \int_\Omega v\, dx$ is clearly a bounded linear operator from $L_1(\Omega)$ into $\mathbb{R}$. Hence, the result is observed by the theorem of differentiation for composed operators. $\qquad\square$

Theorem 1.16 *Let* $f \in \mathcal{C}^3(\mathbb{R})$. *For* $n/2 < p < \infty$, *the operator* $u \mapsto f(u)$ *from* $H_p^2(\Omega)$ *into itself is continuously Fréchet differentiable with the derivative* $h \mapsto f'(u)h$.

Proof We notice that $H_p^2(\Omega) \subset \mathcal{C}(\overline{\Omega})$. Then, the proof is carried out analogously to that of [Yag10, Chapter 1, Subsection 11.10, (5)]. $\qquad\square$

Theorem 1.17 *Let* $f \in \mathcal{C}^{m+1}(\mathbb{R})$ *with some integer* m. *For* $1 \leq s$, $n/2 < s \leq m$, *the operator* $u \mapsto f(u)$ *from* $H^s(\Omega)$ *into itself is continuously Fréchet differentiable with the derivative* $h \mapsto f'(u)h$.

Proof The proof is carried out analogously to that of [Yag10, Chapter 1, Subsection 11.10, (6)]. $\qquad\square$

1.5.2 Gâteaux Differentiation

Let us now verify Gâteaux differentiability for the operator $u \mapsto f(x, u)$, where $f(x, u) \in \mathcal{C}_\infty^{0,1}(\overline{\Omega} \times \mathbb{R})$, from $L_p(\Omega)$ into itself.

Theorem 1.18 *Let* $f \in \mathcal{C}_\infty^{0,1}(\overline{\Omega} \times \mathbb{R})$. *For* $1 \leq p < \infty$, *the operator* $u \mapsto f(x, u)$ *from* $L_p(\Omega)$ *into itself is Gâteaux differentiable with the derivative* $B(u, h) = f_u(x, u)h$. *The derivative is a bounded linear operator of* $L_p(\Omega)$ *and is strongly continuous, that is, for each fixed* $h \in L_p(\Omega)$, *as* $u \to \overline{u}$ *in* $L_p(\Omega)$, $B(u, h) \to B(\overline{u}, h)$ *in* $L_p(\Omega)$.

Proof Let $u, h \in L_p(\Omega)$. For a.e. $x \in \Omega$, (1.54) gives that

$$f(x, u(x) + \theta h(x)) - f(x, u(x)) - \theta f_u(x, u(x))h(x)$$
$$= \theta \int_0^1 [f_u(x, u(x) + \vartheta\theta h(x)) - f_u(x, u(x))]h(x)d\vartheta.$$

Therefore,

$$|f(x, u(x) + \theta h(x)) - f(x, u(x)) - \theta f_u(x, u(x))h(x)|^p$$

$$\leq \theta^p \int_0^1 |[f_u(x, u(x) + \vartheta\theta h(x)) - f_u(x, u(x))]h(x)|^p \, d\vartheta.$$

Integration in Ω yields that

$$\|f(x, u + \theta h) - f(x, u) - \theta f_u(x, u)h\|_{L_p}^p$$

$$\leq \theta^p \int_\Omega \int_0^1 |[f_u(x, u(x) + \vartheta\theta h(x)) - f_u(x, u(x))]h(x)|^p \, dx d\vartheta.$$

Here, as $\theta \to 0$, $[f_u(x, u(x) + \vartheta\theta h(x)) - f_u(x, u(x))]$ converges to 0 for a.e. $(x, \vartheta) \in \Omega \times (0, 1)$. So, the dominated convergence theorem provides that

$$\text{as } \theta \to 0, \quad \|\theta^{-1}[f(x, u + \theta h) - f(x, u)] - f_u(x, u)h\|_{L_p} \to 0.$$

This shows that the condition (1.53) is fulfilled and its derivative is given by $G(u, h) = f_u(x, u)h$.

In view of $f_u(x, u) \in \mathcal{C}_\infty^{0,0}(\overline{\Omega} \times \mathbb{R})$, the derivative is seen to be a bounded linear operator of $L_p(\Omega)$ for h.

Next, let us verify strong continuity of the Gâteaux derivative. Suppose that this would not be strongly continuous. Then, there must exist some $h \in L_p(\Omega)$ and a sequence u_n tending to $\overline{u}$ in $L_p(\Omega)$ such that $\|[f_u(x, u_n) - f_u(x, \overline{u})]h\|_{L_p} \geq \varepsilon$ with some constant $\varepsilon > 0$. As $|u_n - \overline{u}|$ converges to 0 in $L_p(\Omega)$, $|u_n - \overline{u}|$ converges to 0 in measure in Ω. So, we can extract a subsequence $u_{n'}$ such that $u_{n'}(x)$ converges to $\overline{u}(x)$ for a.e. $x \in \Omega$. Then, the dominated convergence theorem again provides that $f_u(x, u_{n'})h$ converges to $f_u(x, \overline{u})h$ in $L_p(\Omega)$, which is a contradiction.   □

According to Theorem 1.14, the operator treated in this theorem is Fréchet differentiable from $L_q(\Omega)$ into $L_p(\Omega)$ if $p < q$. But, when $p = q$, we can no longer prove its Fréchet differentiability (see Remark 1.3).

Corollary 1.2 *Let $f \in \mathcal{C}_\infty^{0,1}(\overline{\Omega} \times \mathbb{R})$. For $1 \leq p < \infty$, if $u \in \mathcal{C}^1([0, T]; L_p(\Omega))$, then $f(x, u(t)) \in \mathcal{C}^1([0, T]; L_p(\Omega))$ with $\frac{d}{dt}f(x, u(t)) = f_u(x, u(t))u'(t)$.*

Proof By virtue of Theorem 1.18, for $u, h \in L_p(\Omega)$ we know that the function $f(x, u + \theta h)$ of θ with values in $L_p(\Omega)$ is a $\mathcal{C}^1$ function. So, we have $f(x, u + h) - f(x, u) = \int_0^1 f_u(x, u + \theta h)h \, d\theta$; here, the integral is taken in the topology of

$L_p(\Omega)$. Use this formula with $u = u(t)$ and $h = u(t + \Delta t) - u(t)$ to obtain that

$$f(x, u(t + \Delta t)) - f(x, u(t))$$

$$= \int_0^1 f_u(x, \theta u(t + \Delta t) + (1 - \theta)u(t))[u(t + \Delta t) - u(t)]d\theta.$$

We can then write

$$\frac{f(x, u(t + \Delta t)) - f(x, u(t))}{\Delta t} - f_u(x, u(t))u'(t)$$

$$= \int_0^1 f_u(x, \theta u(t + \Delta t) + (1 - \theta)u(t)) \left(\frac{u(t + \Delta t) - u(t)}{\Delta t} - u'(t) \right) d\theta$$

$$+ \int_0^1 [f_u(x, \theta u(t + \Delta t) + (1 - \theta)u(t)) - f_u(x, u(t))]u'(t) \, d\theta.$$

As $\Delta t \to 0$, the first integral in the right hand side clearly tends to 0. Meanwhile, the second integral also tends to 0 because of the strong continuity observed above of multiplicative operator by the function $f_u(x, u(x))$ for any $u \in L_p(\Omega)$. Therefore, it is proved that $\frac{d}{dt} f(x, u(t)) = f_u(x, u(t))u'(t)$ for any $0 \le t \le T$.

Continuity of $f_u(x, u(t))u'(t)$ as an $L_p(\Omega)$-valued function of t is similarly shown by the strong continuity of multiplicative operators. □

Corollary 1.3 *Let* $f \in \mathcal{C}_\infty^{0,1}(\overline{\Omega} \times \mathbb{R})$. *If* $u \in \mathcal{C}^1([0, T]; L_1(\Omega))$, *then it holds that* $\int_\Omega f(x, u(t))dx \in \mathcal{C}^1([0, T])$ *with* $\frac{d}{dt} \int_\Omega f(x, u(t))dx = \int_\Omega f_u(x, u(t))u'(t)dx$.

Proof The function $\int_\Omega f(x, u(t))dx$ of t is considered as a composition of a function $t \mapsto f(x, u(t))$ with values in $L_1(\Omega)$ and a continuous linear functional $\varphi \mapsto \int_\Omega \varphi(x)dx$ on $L_1(\Omega)$. Then, the results are obtained directly from the preceding corollary and the fact that the bounded linear operator has its Fréchet derivative coinciding with itself. □

1.6 Some Other Materials

We will collect some other materials we need in this monograph.

1.6.1 Fredholm Operators

Let X and Y be two real Banach spaces. A bounded linear operator T from X into Y is called *a Fredholm operator* if its kernel $\mathcal{K}(T)$ is a subspace of X of finite dimension and its range $\mathcal{R}(T)$ is a closed subspace of Y of finite codimension.

The following theorem in the Riesz–Schauder theory gives an important class of Fredholm operators.

Theorem 1.19 *Let K be a compact linear operator from X into itself. Then, $T = I - K$ is a Fredholm operator, where I denotes the identity operator of X. Moreover, it holds that* $\dim \mathcal{K}(T) = \operatorname{codim} \mathcal{R}(T)$.

For the proof of the theorem, see [DL84a, Lemma II.4.3] or [Bre11, Theorem 6.6].

As an immediate consequence of this theorem, we have the following corollary.

Corollary 1.4 *Let a bounded linear operator T from X into Y be given as $T = S - K$, where S is an isomorphism from X onto Y and K is a compact operator from X into Y. Then, T is a Fredholm operator with* $\dim \mathcal{K}(T) = \operatorname{codim} \mathcal{R}(T)$.

1.6.2 Łojasiewicz Gradient Inequality

Let Ω denote a domain of $\mathbb{R}^N$. Consider a real-valued smooth function $\phi(\xi)$ for $\xi = (\xi_1, \ldots, \xi_N) \in \Omega$ and let $\bar{\xi} \in \Omega$ be its critical point, i.e., $\nabla \phi(\bar{\xi}) = 0$. So, the surface $\xi_{N+1} = \phi(\xi)$ in the space $\mathbb{R}^N \times \mathbb{R}$ is tangential to the plane $\xi_{N+1} = \phi(\bar{\xi})$ at $\bar{\xi}$. When $\phi(\xi)$ is analytic in Ω, the norm of $\nabla \phi(\xi)$ can be estimated from below by a power function of $|\phi(\xi) - \phi(\bar{\xi})|$ in a neighborhood of $\bar{\xi}$.

Theorem 1.20 *Let $\phi : \Omega \to \mathbb{R}$ be an analytic function defined in a domain $\Omega \subset \mathbb{R}^N$ and let $\bar{\xi} \in \Omega$ be its critical point. Then, there exist an exponent $0 < \theta \leq \frac{1}{2}$ and a neighborhood $U \subset \Omega$ of $\bar{\xi}$ such that the inequality*

$$\|\nabla \phi(\xi)\|_{\mathbb{R}^N} \geq D|\phi(\xi) - \phi(\bar{\xi})|^{1-\theta}, \qquad \xi \in U, \tag{1.55}$$

holds true with some constant $D > 0$.

This inequality was first presented by Łojasiewicz [Loj63, Loj65]. So, (1.55) is called *the Łojasiewicz gradient inequality*.

1.7 Notes

The basic materials of Complex Functional Analysis presented in Section 1.1 were all explained or obtained in [Yag10]. As for dual spaces and adjoint spaces, see Subsections 6.1 and 6.2 of [Yag10, Chapter 1] respectively. Interpolation spaces are explained in Subsection 5.1 of [Yag10, Chapter 1], see also [Tri78, Section 1.9]. Triplets of spaces are constructed in Subsection 7.1 of [Yag10, Chapter 1]. Sectorial operators are defined and studied in [Yag10, Chapter 2]. Theorem 1.4 is verified by Theorems 4.1 and 4.4 in [Yag10, Chapter 4].

The notion of conjugations in complex Banach spaces and the notion of real operators acting in conjugated Banach spaces were originally presented in the paper [Yag17]. These various properties explained in Sections 1.2 and 1.3 were also obtained there. As those are relatively new and give us important frameworks of study, we presented them again here together with their full proofs.

Let T be a compact operator of a Banach space. By virtue of the Riesz–Schauder theory, we can know the structure of its spectrum $\sigma(T)$ or the necessary and sufficient condition for solvability of the vector equation $Tu = f$, f being a given vector and u being an unknown vector, in terms of its dual operator T'. Theorem 1.19 is then shown in that theory and plays an important role.

Except for the case where $N = 1$, the proof of the gradient inequality (1.55) is not at all elementary. As for its proof, see also Łojasiewicz–Zurro [LZ99].

Chapter 2
Review of Abstract Results

This chapter is devoted to reviewing the abstract results in [Yag].

2.1 General Settings

Consider the following general frameworks, as in Section 2.2 of [Yag].

Let X be a real Hilbert space with inner product $(\cdot, \cdot)_X$ and norm $\| \cdot \|_X$, and let Z be a second real Hilbert space with norm $\| \cdot \|_Z$, Z being embedded in X densely and continuously. According to Definition 1.2, we introduce the triplet

$$Z \subset X \subset Z^* \tag{2.1}$$

of real spaces. The norm of Z^* is denoted by $\| \cdot \|_{Z^*}$ and the duality product between Z and Z^* is denoted by $\langle \cdot, \cdot \rangle_{Z \times Z^*}$. Then, $(u, f)_X = \langle u, f \rangle_{Z \times Z^*}$ holds for $u \in Z$ and $f \in X$ due to (1.8). By the property (4) of Theorem 1.10, we have

$$\|u\|_X \leq C \|u\|_Z^{\frac{1}{2}} \|u\|_{Z^*}^{\frac{1}{2}}, \qquad u \in Z, \tag{2.2}$$

with some constant $C > 0$.

Let $u(t)$ be a function defined on $[0, \infty)$ which lies in the function space

$$u \in \mathcal{C}([0, \infty); Z) \cap \mathcal{C}^1([0, \infty); Z^*). \tag{2.3}$$

Assume that $u(t)$ satisfies a global norm estimate

$$\|u(t)\|_Z \leq R \quad \text{for all} \quad 0 \leq t < \infty, \tag{2.4}$$

with some constant $R > 0$.

© The Author(s), under exclusive license to Springer Nature Singapore Pte Ltd. 2021
A. Yagi, *Abstract Parabolic Evolution Equations and Łojasiewicz–Simon Inequality II*,
SpringerBriefs in Mathematics, https://doi.org/10.1007/978-981-16-2663-0_2

Meanwhile, let us introduce a functional $\Phi(f)$ defined for $f \in X$ with values in $\mathbb{R}$. Assume that this functional is continuously differentiable in the sense that, for each $f \in X$, there exists a vector $\dot{\Phi}(f) \in X$ for which it holds that

$$\text{as } h \to 0 \text{ in } X, \quad \Phi(f + h) - \Phi(f) - (\dot{\Phi}(f), h)_X = o(\|h\|_X), \tag{2.5}$$

and that

$$\text{the mapping } f \mapsto \dot{\Phi}(f) \text{ is continuous from } X \text{ into itself.} \tag{2.6}$$

As a matter of fact, (2.5)–(2.6) are equivalent to the condition:

$$\Phi : X \to \mathbb{R} \text{ is continuously Fréchet differentiable.} \tag{2.7}$$

Indeed, the isometric isomorphism $J : X \to X'$ presented in (1.35) gives the relation. Therefore, (2.7) is certainly equivalent to (2.5)–(2.6).

Furthermore, assume that, when $u \in Z$, the derivative $\dot{\Phi}(u)$ is in Z and that

$$\text{the mapping } u \mapsto \dot{\Phi}(u) \text{ is continuous from } Z \text{ into itself.} \tag{2.8}$$

Similarly, this is implied by the conditions that, when $u \in Z$, the derivative $\Phi'(u)$ is in $(Z^*)' \subset X'$ and that

$$\text{the mapping } u \mapsto \Phi'(u) \text{ is continuous from } Z \text{ into } (Z^*)'. \tag{2.9}$$

Indeed, the operator J in (1.35) is an isometric isomorphism from Z onto $(Z^*)'$.

Under the situation above, the following proposition is verified, see [Yag, Proposition 2.3].

Proposition 2.1 *The function $\Phi(u(t))$ is continuously differentiable for $0 \leq t < \infty$ and the derivative is given by*

$$\frac{d}{dt}\Phi(u(t)) = \left\langle \dot{\Phi}(u(t)), u'(t) \right\rangle_{Z \times Z^*}, \qquad 0 \leq t < \infty. \tag{2.10}$$

As usual, the ω-limit set of the function $u(t)$ is defined by

$$\omega(u) = \{\bar{u} \in Z^*; \ \exists t_n \nearrow \infty \text{ such that } u(t_n) \to \bar{u} \text{ in } Z^*\}. \tag{2.11}$$

As the closed ball $\overline{B}^Z(0; R)$ is sequentially weakly compact in Z, we can assume that, if $\overline{u} \in \omega(u)$, then $u(t_n) \to \overline{u}$ weakly in Z; in particular, $\overline{u} \in \overline{B}^Z(0; R)$. Furthermore, by (2.2) and (2.4), $u(t_n) \to \overline{u}$ in Z^* implies $u(t_n) \to \overline{u}$ in X, too. Therefore, $\lim_{n\to\infty} \Phi(u(t_n)) = \Phi(\overline{u})$.

2.2 Structural Assumptions and Main Theorem

Make the following structural assumptions, as in Section 2.3 of [Yag].

There exists an ω-limit $\overline{u}$ of $u(t)$ satisfying the following four conditions:

(I) *Critical Condition.* The $\overline{u}$ is a critical point of $\Phi(u)$, i.e., $\Phi'(\overline{u}) = 0$ (which is equivalent to $\dot{\Phi}(\overline{u}) = 0$).

(II) *Lyapunov Function.* There exists a radius $r' > 0$ such that

$$\Phi(u(t)) > \Phi(\overline{u}) \quad \text{and} \quad \frac{d}{dt}\Phi(u(t)) \leq 0 \quad \text{if } u(t) \in B^X(\overline{u}; r'). \quad (2.12)$$

(III) *Angle Condition.* There exist a radius $r'' > 0$ and a constant $\delta > 0$ such that

$$-\langle \dot{\Phi}(u(t)), u'(t) \rangle_{Z \times Z^*} \geq \delta \|\dot{\Phi}(u(t))\|_Z \|u'(t)\|_{Z^*} \quad \text{if } u(t) \in B^X(\overline{u}; r''). \quad (2.13)$$

(IV) *Gradient Inequality.* There exist a radius $r''' > 0$ and an exponent $0 < \theta \leq \frac{1}{2}$ for which it holds that

$$\|\dot{\Phi}(u(t))\|_Z \geq D|\Phi(u(t)) - \Phi(\overline{u})|^{1-\theta} \quad \text{if } u(t) \in B^X(\overline{u}; r'''), \quad (2.14)$$

$D > 0$ being some constant.

Then, the following result is proved, see [Yag, Theorem 2.1].

Theorem 2.1 *Under the situations* (2.1), (2.3), (2.4), (2.5), (2.6) *and* (2.8), *let an ω-limit $\overline{u}$ satisfy the structural conditions* (I), (II), (III) *and* (IV). *Then, as $t \to \infty$, $u(t)$ converges to $\overline{u}$ at the rate*

$$\|u(t) - \overline{u}\|_{Z^*} \leq (D\delta\theta)^{-1}[\Phi(u(t)) - \Phi(\overline{u})]^\theta \quad \text{for all } t \geq T, \quad (2.15)$$

where $T > 0$ is some time which is fixed sufficiently large.

2.3   Łojasiewicz–Simon Gradient Inequality

Among the structural assumptions (I)–(IV), the most important one is the gradient inequality (2.14) which is called the Łojasiewicz–Simon Gradient Inequality. In this section, let us review the method to obtain (2.14) in the present general settings.

2.3.1   Settings and Structural Assumptions

Let X be a real Hilbert space with inner product $(\cdot, \cdot)_X$. Consider a functional $\Phi : X \to \mathbb{R}$ which is continuously differentiable in the sense of (2.5)–(2.6) with derivative $\dot{\Phi} : X \to X$. Let $\overline{u} \in X$ be a critical point of $\Phi(u)$, i.e.,

$$\Phi'(\overline{u}) = J\dot{\Phi}(\overline{u}) = 0. \tag{2.16}$$

We first assume that

$$\dot{\Phi} : X \to X \text{ is Gâteaux differentiable at } \overline{u} \tag{2.17}$$

and that the derivative $L = [\dot{\Phi}]'(\overline{u})$ is not only a bounded linear operator from X into itself but also

$$L = [\dot{\Phi}]'(\overline{u}) \text{ is a Fredholm operator of } X. \tag{2.18}$$

It is possible to read (2.17)–(2.18) as conditions for the Fréchet derivative $\Phi' : X \to X'$. Because of $\dot{\Phi} = J^{-1}\Phi'$, where $J : X \to X'$ is the isometric isomorphism, (2.17) is equivalent to the Gâteaux differentiability of $\Phi'(u)$ at $\overline{u}$. In addition, (2.18) is equivalent to saying that its derivative $[\Phi']'(\overline{u})$ is a bounded linear operator from X into X' and is a Fredholm operator.

Let us denote the kernel (resp. range) of L by $\mathcal{K}(L)$ (resp. $\mathcal{R}(L)$). By definition, $\mathcal{K}(L)$ is a finite-dimensional space of X and $\mathcal{R}(L)$ is a closed subspace of X having a finite codimension.

We next assume that there exists a real Banach space Y which is a linear subspace of X and satisfies the following five conditions:

$$Y \subset X \text{ with dense and continuous embedding;} \tag{2.19}$$

$$\overline{u} \in Y; \tag{2.20}$$

$$L^{-1}(Y) \subset Y, \text{ i.e., if } Lu \in Y, \text{ then } u \in Y; \tag{2.21}$$

$$\dot{\Phi} \text{ maps } Y \text{ into itself;} \tag{2.22}$$

$$\dot{\Phi} : Y \to Y \text{ is continuously Fréchet differentiable in } Y. \tag{2.23}$$

2.3.2 Some Properties of L

We can then verify the following properties for the operator L in X. First, L is symmetric in X, that is,

$$(Lu, v)_X = (u, Lv)_X \quad \text{for any } u, v \in X. \tag{2.24}$$

As $\mathcal{K}(L)$ is finite-dimensional, $\mathcal{K}(L)$ is a closed subspace of X. Denote by P the orthogonal projection from X onto $\mathcal{K}(L)$. By definition, we have $LP = 0$ on X. But, since P is also symmetric in X, it follows that $PL = (LP)^* = 0$, i.e.,

$$LP = PL = 0 \quad \text{on} \quad X. \tag{2.25}$$

The projection P naturally induces an orthogonal decomposition of X into $X = PX + (I - P)X$, where $PX = (I - P)^{-1}0 = \mathcal{K}(L)$ and $(I - P)X = P^{-1}0$. This means that L yields an orthogonal decomposition of X of the form

$$X = \mathcal{K}(L) + L(X), \tag{2.26}$$

its projections being given by P and $I - P$ respectively. Consequently, L is an isomorphism from $L(X)$ onto itself.

Since $\dot{\Phi}$ is Gâteaux differentiable at $\overline{u}$ in X and since $\dot{\Phi}$ is Fréchet differentiable in Y, we naturally have $[\dot{\Phi}]'(\overline{u})h = Lh$ for $h \in Y$. Because of $[\dot{\Phi}]'(\overline{u}) \in \mathcal{L}(Y)$, this means that L is a bounded linear operator of Y, too.

Notice that (2.21) implies that $\mathcal{K}(L) = L^{-1}0 \subset L^{-1}Y \subset Y$. Therefore, $\mathcal{K}(L)$ is also a finite-dimensional subspace of Y. In particular, $\mathcal{K}(L)$ is a closed subspace of Y, too. In addition, since the two norms $\| \cdot \|_Y$ and $\| \cdot \|_X$ are equivalent on the space $\mathcal{K}(L)$, we see that $\|Pu\|_Y \leq C\|Pu\|_X \leq C\|u\|_X$, which means that P is a bounded operator from Y into itself. Therefore, we have $P \in \mathcal{L}(Y)$ and $P^2 = P$. Consequently, P induces a topological decomposition of Y, too, in the manner that $Y = PY + (I - P)Y$. Since $PY = (I - P)^{-1}0$ and $(I - P)Y = P^{-1}0$, PY and $(I - P)Y$ are a closed subspace of Y. Hence, L yields a topological decomposition of Y, too, in the form

$$Y = \mathcal{K}(L) + L(Y). \tag{2.27}$$

Moreover, L is an isomorphism from $L(Y)$ onto itself.

2.3.3 Critical Manifold

As in Chill [Chi06], we are led to consider the surface

$$S = \{u \in Y; \ \dot{\Phi}(u) \in \mathcal{K}(L)\} = \{u \in Y; \ (I - P)\dot{\Phi}(u) = 0\} \tag{2.28}$$

of Y, which is called the critical manifold of $\Phi(u)$.

In view of (2.27), it is possible to identify Y as the product space $Y = \mathcal{K}(L) \times L(Y)$. Then, in a suitable neighborhood of $\bar{u}$, S can be represented as follows. There is an open neighborhood $U = U_0 \times U_1$ of $\bar{u}$ in Y, where U_0 (resp. U_1) is an open neighborhood of $P\bar{u}$ (resp. $(I - P)\bar{u}$) in $\mathcal{K}(L)$ (resp. $L(Y)$), such that S is given in U by

$$S \cap U = \{(u_0, g(u_0)); \ u_0 \in U_0, \ g : U_0 \to U_1\},$$

where g is a mapping from U_0 into U_1 satisfying $g(P\bar{u}) = (I - P)\bar{u}$ and is continuously Fréchet differentiable.

Using the mapping $g : U_0 \to U_1$, we can introduce another decomposition for the vectors of U (instead of (2.27)). In fact, for $u \in U$, u can be expressed in the form

$$u = [Pu + g(Pu)] + [(I - P)u - g(Pu)] \equiv u_S + u_1 \in S + L(Y), \qquad u \in U, \tag{2.29}$$

where $u_S = Pu + g(Pu) = (Pu, g(Pu)) \in S$ and $u_1 = (I - P)u - g(Pu) \in L(Y)$ respectively. We also notice that

$$\text{as } u \to \bar{u} \text{ in } Y, u_S \text{ (resp. } u_1 \text{) converges to } \bar{u} \text{ (resp. 0) in } Y. \tag{2.30}$$

This decomposition then provides the following important estimate for $\dot{\Phi}(u)$.

Proposition 2.2 *It holds true that*

$$\|P\dot{\Phi}(u)\|_Y \geq \|\dot{\Phi}(u_S)\|_Y - o(1)\|u_1\|_Y, \qquad u = u_S + u_1 \in U, \tag{2.31}$$

$$\|(I - P)\dot{\Phi}(u)\|_Y \geq c\|u_1\|_Y, \qquad u = u_S + u_1 \in U, \tag{2.32}$$

provided that U is a sufficiently small neighborhood of $\bar{u}$ in Y. Here, $o(1)$ denotes a small quantity tending to 0 as $u \to \bar{u}$ in Y, while c is a fixed positive constant.

Combining (2.31) and (2.32), we verify that

$$\|\dot{\Phi}(u)\|_Y \geq C\big[\|P\dot{\Phi}(u)\|_Y + \|(I - P)\dot{\Phi}(u)\|_Y\big]$$
$$\geq C\big[\|\dot{\Phi}(u_S)\|_Y + (c - o(1))\|u_1\|_Y\big].$$

Hence, by virtue of (2.30), it follows that

$$\|\dot{\Phi}(u)\|_Y \geq c[\|\dot{\Phi}(u_S)\|_Y + \|u_1\|_Y], \qquad u = u_S + u_1 \in U, \tag{2.33}$$

provided that U is a sufficiently small neighborhood of $\bar{u}$ in Y, where c is a fixed positive constant.

2.3.4 Main Theorem

Utilizing the decomposition (2.29) and the estimate (2.33), we want to estimate the difference $|\Phi(u) - \Phi(\overline{u})|$, writing it as

$$|\Phi(u) - \Phi(\overline{u})| \leq |\Phi(u) - \Phi(u_S)| + |\Phi(u_S) - \Phi(\overline{u})|, \qquad u = u_S + u_1 \in U.$$

First, by the assumption (2.23), the former difference can be shown to be such that

$$|\Phi(u) - \Phi(u_S)| \leq \|\dot{\Phi}(u_S)\|_X^2 + C\|u_1\|_Y^2.$$

Therefore, by (2.33), we obtain that

$$\|\dot{\Phi}(u)\|_Y \geq C|\Phi(u) - \Phi(u_S)|^{\frac{1}{2}}, \qquad \forall u \in U, \tag{2.34}$$

where U is the same neighborhood as in (2.33).

Second, we notice that Proposition 2.2 implies that u_S is described as $u_S = u_0 + g(u_0)$, where $u_0 \in U_0 \subset \mathcal{K}(L)$ and $g(u_0) \in U_1 \subset L(Y)$. So, let $v_1, v_2, \ldots, v_N$ be a basis of $\mathcal{K}(L)$, where $N = \dim \mathcal{K}(L)$, and identify $\mathcal{K}(L)$ with $\mathbb{R}^N$ by the correspondence

$$u_0 = \sum_{k=1}^{N} \xi_k v_k \in \mathcal{K}(L) \quad \longleftrightarrow \quad \boldsymbol{\xi} = (\xi_1, \xi_2, \ldots, \xi_N) \in \mathbb{R}^N.$$

Let $P\overline{u} \leftrightarrow \overline{\boldsymbol{\xi}}$ and let U_0 correspond to an open neighborhood Ω of $\overline{\boldsymbol{\xi}}$ in $\mathbb{R}^N$. Here we make the crucial assumption that

the function $\boldsymbol{\xi} \in \Omega \mapsto \phi(\boldsymbol{\xi}) \equiv \Phi\left(\sum_{k=1}^{N} \xi_k v_k + g(\sum_{k=1}^{N} \xi_k v_k)\right)$

is analytic in a neighborhood of $\overline{\boldsymbol{\xi}}$. (2.35)

Then, the classical Theorem 1.20 provides that, for some exponent $0 < \theta \leq \frac{1}{2}, \phi(\boldsymbol{\xi})$ satisfies the gradient inequality

$$\|\nabla_{\boldsymbol{\xi}}\phi(\boldsymbol{\xi})\|_{\mathbb{R}^N} \geq D_0|\phi(\boldsymbol{\xi}) - \phi(\overline{\boldsymbol{\xi}})|^{1-\theta}, \qquad \boldsymbol{\xi} \in \Omega,$$

with some constant $D_0 > 0$, provided that Ω is replaced by a smaller one. Furthermore, we observe from this that

$$\|\dot{\Phi}(u_S)\|_X \geq C|\Phi(u_S) - \Phi(\overline{u})|^{1-\theta}, \qquad u_S = u_0 + g(u_0), \ u_0 \in U_0,$$

provided that U_0 is replaced by a smaller one.

According to (2.33), $\|\dot{\Phi}(u)\|_Y \geq C\|\dot{\Phi}(u_S)\|_Y$ for $u \in U$ if U is a sufficiently small neighborhood of $\overline{u}$ in Y. Hence, it follows that

$$\|\dot{\Phi}(u)\|_Y \geq C|\Phi(u_S) - \Phi(\overline{u})|^{1-\theta}, \qquad \forall u \in U. \tag{2.36}$$

In this way, we can obtain the following theorem.

Theorem 2.2 *Let* $\Phi : X \rightarrow \mathbb{R}$ *be a continuously differentiable functional in the sense of* (2.5)–(2.6) *with derivative* $\dot{\Phi} : X \rightarrow X$, *and let* $\overline{u}$ *be its critical point (as* (2.16)). *Assume that* $u \mapsto \dot{\Phi}(u)$ *is Gâteaux differentiable at* $\overline{u}$ *(as* (2.17)) *with derivative* L *satisfying* (2.18). *Assume also that there exists a Banach space* Y *satisfying* (2.19)–(2.23). *Define the critical manifold* S *as* (2.28) *and assume that the analyticity condition* (2.35) *is satisfied. Then, in a neighborhood* U *of* $\overline{u}$ *in* Y, $\Phi(u)$ *satisfies the gradient inequality*

$$\|\dot{\Phi}(u)\|_Y \geq C|\Phi(u) - \Phi(\overline{u})|^{1-\theta}, \qquad \forall u \in U, \tag{2.37}$$

where $0 < \theta \leq \frac{1}{2}$ *is the exponent given in* (2.36).

2.3.5   Gradient Inequality with Respect to $\|\cdot\|_Z$

What we have to prove is the gradient inequality (2.14) with respect to the norm of Z. Therefore, it still remains to show the way how to derive (2.14) from (2.37).

(I) *Case where* $Z \subset Y$. Consider first the case where Y includes Z. More precisely, Y is such that $Z \subsetneq Y \subset X$ and that the intermediate inequality

$$\|u\|_Y \leq C\|u\|_Z^\alpha \|u\|_X^{1-\alpha}, \qquad u \in Z, \tag{2.38}$$

holds with some $0 \leq \alpha < 1$. Let U be the neighborhood of $\overline{u}$ in Y stated in Theorem 2.2. Then, there exists $r''' > 0$ for which we have $B^X(\overline{u}; r''') \cap \overline{B}^Z(0; R) \subset U$, where R is the radius appearing in (2.4). Since $\|\dot{\Phi}(u)\|_Z \geq C\|\dot{\Phi}(u)\|_Y$ for $u \in Z$, (2.37) can yield the desired inequality (2.14).

(II) *Case where* $Y \subset Z$. In this case, (2.37) cannot yield (2.14) in any direct way. So, we need to introduce another supplementary Banach space $W \subset Y$ and to assume a global uniform boundedness of $u(t)$ with respect to the W-norm.

More precisely, let W denote a Banach space such that $W \subset Y \subset Z$. Assume that, among the spaces $W \subset Y \subset X$, the intermediate inequality

$$\|u\|_Y \leq C\|u\|_W^\alpha \|u\|_X^{1-\alpha}, \qquad u \in W, \tag{2.39}$$

holds with some $0 < \alpha < 1$ and that, among the spaces $W \subset Y \subset Z$, the estimate

$$\|u\|_Y \leq C \|u\|_W^\beta \|u\|_Z^{1-\beta}, \qquad u \in W, \tag{2.40}$$

holds with some $0 \leq \beta < \theta$, θ being the exponent appearing in (2.37).

Meanwhile, assume that the solution $u(t)$ satisfies the global norm estimate

$$\|u(t)\|_W \leq R', \qquad 0 \leq t < \infty, \tag{2.41}$$

with some constant $R' > 0$ and that $u \mapsto \dot{\Phi}(u)$ is a mapping from W into itself together with the estimate

$$\|\dot{\Phi}(u)\|_W \leq C, \qquad u \in \overline{B}^W(0; R'). \tag{2.42}$$

It then follows by (2.39) that there exists $r''' > 0$ such that

$$B^X(\overline{u}; r''') \cap \overline{B}^W(0; R') \subset U,$$

where U is the neighborhood of $\overline{u}$ in Y stated in Theorem 2.2. On the other hand, by (2.40) and (2.42), we have

$$C \|\dot{\Phi}(u)\|_Z^{1-\beta} \geq \|\dot{\Phi}(u)\|_Y, \qquad u \in \overline{B}^W(0; R').$$

Therefore, (2.37) yields the estimate

$$\|\dot{\Phi}(u)\|_Z \geq D' |\Phi(u) - \Phi(\overline{u})|^{1-\theta'}, \qquad u \in B^X(\overline{u}; r''') \cap \overline{B}^W(0; R'), \tag{2.43}$$

with an exponent $\theta' = \frac{\theta - \beta}{1 - \beta}$ which is positive because of $\beta < \theta$.

Thus, (2.43) together with (2.41) can yield the desired inequality (2.14).

2.4 Comments for Applications

In this chapter, we have introduced the general settings of study, have made the structural assumptions and have proved the convergence theorem. It may be meaningful here to explain how to choose the triplet $Z \subset X \subset Z^*$ and how to verify the four assumptions (I)–(IV) in a general context.

We suppose that the function $u(t)$ is usually given as a global solution of some evolution equation $u' = F(u)$, $0 \leq t < \infty$, where $F(u)$ is a continuous nonlinear operator from a Hilbert space Z into its adjoint space Z^*, Z being compactly embedded in Z^*. Therefore, $u(t)$ is naturally set as a function belonging to $\mathcal{C}([0, \infty); Z) \cap \mathcal{C}^1([0, \infty); Z^*)$ together with the boundedness (2.4). Existence of a Lyapunov function $\Phi(u)$ defined in a Hilbert space X, where the spaces

$Z \subset X \subset Z^*$ make a triplet, is a major premise of our arguments. Along the trajectory of $u(t)$, it holds that $\frac{d}{dt}\Phi(u(t)) \leq 0$.

(I) *Critical Condition.* As (3.46), integration of $\frac{d}{dt}\Phi(u(t)) \leq 0$ for $t \in [0, \infty)$ yields integrability $\int_0^\infty \|u'(t)\|_{Z^*}^2 dt < \infty$. Therefore, there exists a temporal sequence $t_n \nearrow \infty$ such that $u'(t_n) \to 0$ in Z^*, i.e., $F(u(t_n)) \to 0$ in Z^*. Meanwhile, due to (2.4), we can assume that $u(t_n) \to \overline{u} \in \omega(u)$ in Z^* and $u(t_n) \to \overline{u}$ weakly in Z, which implies that $F(u(t_n)) \to F(\overline{u})$ weakly in Z^*. Hence, we have $F(\overline{u}) = 0$. But, it is in general the case that $\dot{\Phi}(u) = 0$ if and only $F(u) = 0$. Thereby, $F(\overline{u}) = 0$ implies $\dot{\Phi}(\overline{u}) = \Phi'(\overline{u}) = 0$.

(II) *Lyapunov Function.* Assume that $\frac{d}{dt}\Phi(u(t)) = 0$ at some $t = \overline{t}$. Then, (2.10) together with the angle condition (2.13) yields $\dot{\Phi}(u(\overline{t})) = 0$ or $u'(\overline{t}) = 0$. As $\dot{\Phi}(u(\overline{t})) = 0$ yields $F(u(\overline{t})) = 0$, $u(\overline{t})$ is in any case a stationary solution and $u(t) \equiv u(\overline{t})$ for all $t \geq \overline{t}$, and the assertion of Theorem 2.1 is trivial. Therefore, it suffices to argue under the condition that $\frac{d}{dt}\Phi(u(t)) < 0$ for any $0 \leq t < \infty$, provided that the angle condition is fulfilled.

As mentioned, existence of function $\Phi(u)$ satisfying $\frac{d}{dt}\Phi(u(t)) \leq 0$ for $0 \leq t < \infty$ is a major premise in our study.

(III) *Angle Condition.* This condition is more essential than the former two. Very roughly speaking, we can consider that the product $\langle \dot{\Phi}(u(t)), u'(t) \rangle_{Z \times Z^*}$ is given by $\|\dot{\Phi}(u(t))\|_Z \|u'(t)\|_{Z^*} \cos \Theta$, where Θ stands for the angle of two vectors $\dot{\Phi}(u(t))$ and $u'(t)$. Then, (2.13) yields $-\cos \Theta \geq \delta$, namely, Θ must be strictly larger that $\frac{\pi}{2}$. As seen, the most favorable case is that the underlying evolution equations are of gradient form, namely, $u'(t) = F(u(t))$ is always in the direction of $-\dot{\Phi}(u(t))$, i.e., $\Theta = \pi$.

Anyway, (2.13) is verified by careful estimates on the derivative $-\frac{d}{dt}\Phi(u(t))$.

(IV) *Gradient Inequality.* As a matter of fact, this is the most crucial assumption. Consider the surface $\Phi = \Phi(u)$, $u \in Z$, in the product space $(u, \Phi) \in Z \times \mathbb{R}$. The critical condition $\Phi'(\overline{u}) = 0$ means that this surface is tangential to the horizontal plane $\Phi = \Phi(\overline{u})$ at $u = \overline{u}$. The gradient inequality then requires that its tangency must be in a moderate degree. As reviewed above, it is necessary for establishing (2.37) to analyze precisely the behavior of $\dot{\Phi}(u) = J^{-1}\Phi'(u)$ in a neighborhood of $\overline{u}$ and to assume the analyticity condition (2.35) on S.

Chapter 3
Parabolic Equations

In this chapter we study semilinear parabolic equation and reaction–diffusion equations. As for semilinear parabolic equations, we treat Eq. (3.1) in an n-dimensional bounded domain Ω. In order to verify the gradient inequality (2.14), however, the techniques suggested in Subsection 2.3.5 require sufficient regularities for the solutions. For this reason, we have to make quite different basic assumptions on the regularity of Ω and on the regularity of functions $a_{ij}(x)$ and $f(x, u)$ in (3.1), depending on the dimension n.

As for reaction–diffusion equations, we treat only a three-dimensional equation (3.61) of specific form. But if we use the techniques described for the semilinear parabolic equations, then it is equally possible to treat general n-dimensional reaction–diffusion equations.

3.1 Semilinear Parabolic Equations

Consider a semilinear parabolic equation

$$\begin{cases} \dfrac{\partial u}{\partial t} - \displaystyle\sum_{i,j=1}^{n} \dfrac{\partial}{\partial x_i}\left(a_{ij}(x)\dfrac{\partial u}{\partial x_j}\right) = f(x, u) & \text{in } \Omega \times (0, \infty), \\ u = 0 & \text{on } \partial\Omega \times (0, \infty), \end{cases} \tag{3.1}$$

in a bounded domain $\Omega \subset \mathbb{R}^n$, where $n = 2, 3, \ldots$, with boundary $\partial\Omega$. The function $u = u(x, t)$ defined for $(x, t) \in \Omega \times [0, \infty)$ is an unknown function satisfying the homogeneous Dirichlet conditions on $\partial\Omega$.

© The Author(s), under exclusive license to Springer Nature Singapore Pte Ltd. 2021
A. Yagi, *Abstract Parabolic Evolution Equations and Łojasiewicz–Simon Inequality II*,
SpringerBriefs in Mathematics, https://doi.org/10.1007/978-981-16-2663-0_3

Depending on the dimension n, we assume the following regularity for Ω:

$$\begin{cases} \text{when } n = 2, \ \partial\Omega \text{ is a Lipschitz boundary;} \\ \text{when } n = 3, \ \Omega \text{ is convex or } \partial\Omega \text{ is of class } \mathcal{C}^2; \\ \text{when } n = 4, \ \partial\Omega \text{ is of class } \mathcal{C}^3; \\ \text{when } n \geq 5, \ \partial\Omega \text{ is of class } \mathcal{C}^\infty; \end{cases} \tag{3.2}$$

respectively. The coefficients $a_{ij}(x)$ are real-valued functions such that

$$a_{ij}(x) \equiv a_{ji}(x) \quad \text{for any pair } 1 \leq i, j \leq n, \tag{3.3}$$

and it holds that

$$\sum_{i,j=1}^{n} a_{ij}(x)\eta_i\eta_j \geq \delta|\eta|^2, \qquad \eta = (\eta_1, \eta_2, \ldots, \eta_n) \in \mathbb{R}^n, \ x \in \Omega, \tag{3.4}$$

with a positive constant δ. Furthermore, we assume that

$$\begin{cases} \text{when } n = 2, \ a_{ij}(x) \in L_\infty(\Omega), \quad i, j = 1, 2; \\ \text{when } n = 3, \ a_{ij}(x) \in \mathcal{C}^1(\overline{\Omega}), \quad i, j = 1, 2, 3; \\ \text{when } n = 4, \ a_{ij}(x) \in \mathcal{C}^2(\overline{\Omega}), \quad i, j = 1, 2, 3, 4; \\ \text{when } n \geq 5, \ a_{ij}(x) \in \mathcal{C}^\infty(\overline{\Omega}), \quad i, j = 1, 2, \ldots, n; \end{cases} \tag{3.5}$$

respectively. The forcing function $f(x, u)$ is defined for $(x, u) \in \overline{\Omega} \times \mathbb{R}$ and takes real values. The function is assumed to have the following regularity:

$$\begin{cases} \text{when } n = 2 \text{ or } 3, \ f(x, u) \in \mathcal{C}^{0,\infty}(\overline{\Omega} \times \mathbb{R}); \\ \text{when } n = 4, \ f(x, u) \in \mathcal{C}^{1,\infty}(\overline{\Omega} \times \mathbb{R}); \\ \text{when } n \geq 5, \ f(x, u) \in \mathcal{C}^{\infty,\infty}(\overline{\Omega} \times \mathbb{R}); \end{cases} \tag{3.6}$$

respectively.

We will handle (3.1) as an abstract evolution equation

$$u' + Au = f(u), \qquad 0 < t < \infty, \tag{3.7}$$

in $L_2(\Omega)$. Here, A is a realization of the elliptic operator $-\sum_{i,j=1}^{n} \frac{\partial}{\partial x_j}\left(a_{ij}(x)\frac{\partial}{\partial x_i}\right)$ in $L_2(\Omega)$ under the boundary conditions $u = 0$ on $\partial\Omega$, and $f(u)$ is a nonlinear operator in $L_2(\Omega)$ defined as $u \mapsto f(x, u)$.

More precisely, A is defined as follows. Consider the bilinear form

$$a(u, v) = \int_\Omega \sum_{i,j=1}^n a_{ij}(x) \frac{\partial u}{\partial x_i} \frac{\partial v}{\partial x_j}\, dx, \qquad u, v \in \overset{\circ}{H}{}^1(\Omega). \qquad (3.8)$$

By (3.5), $a(u, v)$ is a continuous form on $\overset{\circ}{H}{}^1(\Omega)$. On the other hand, by (3.4), $a(u, u)$ satisfies $a(u, u) \geq \delta \|\nabla u\|_{L_2}^2$ for $u \in \overset{\circ}{H}{}^1(\Omega)$. Since the Poincaré inequality provides that $\|\nabla u\|_{L_2} \geq c\|u\|_{H^1}$ for any $u \in \overset{\circ}{H}{}^1(\Omega)$ with some constant $c > 0$, we obtain that $a(u, u) \geq \delta c^2 \|u\|_{H^1}^2$, which means that $a(u, v)$ is a coercive form on $\overset{\circ}{H}{}^1(\Omega)$. Thereby, $a(u, v)$ is a real sesquilinear form defined in Subsection 1.4.1. Then, by the same procedure as in Subsection 1.4.2, the form $a(u, v)$ determines a densely defined, closed linear operator A of $L_2(\Omega)$ which is a positive definite self-adjoint operator and is a real sectorial operator.

Under (3.2) and (3.5), we can know the domain $\mathcal{D}(A)$ of A by using the results of [Yag10, Chapter 1, Section 1]. When $n = 2$, there is an exponent $p > 2$ such that

$$\mathcal{D}(A) \subset H_p^1(\Omega) \cap \overset{\circ}{H}{}^1(\Omega) \quad \text{(with continuous embedding)}. \qquad (3.9)$$

When $n \geq 3$, $\mathcal{D}(A)$ is given by

$$\mathcal{D}(A) = H^2(\Omega) \cap \overset{\circ}{H}{}^1(\Omega) \quad \text{(with norm equivalence)}. \qquad (3.10)$$

Furthermore, for every n, the domain of its square root is given by

$$\begin{cases} \mathcal{D}(A^{\frac{1}{2}}) = \overset{\circ}{H}{}^1(\Omega) \quad \text{(with norm equivalence)}, \\[2mm] a(u, v) = (A^{\frac{1}{2}}u, A^{\frac{1}{2}}v)_{L_2} \quad \text{for } u, v \in \mathcal{D}(A^{\frac{1}{2}}). \end{cases} \qquad (3.11)$$

Meanwhile, the nonlinear operator $f(u)$ is precisely defined as

$$f(u) = f(x, u) \qquad \text{for } u \in \mathcal{D}(f) \equiv \{u \in L_2(\Omega);\ f(x, u) \in L_2(\Omega)\}.$$

Now, as the prerequisite assumption, let there exist a global solution $u(t)$ to (3.7) in the function space

$$u \in \mathcal{C}([0, \infty); \mathcal{D}(A)) \cap \mathcal{C}^1([0, \infty); L_2(\Omega)) \cap \mathcal{B}([0, \infty); L_\infty(\Omega)). \qquad (3.12)$$

Here we make a crucial assumption that there exists an open interval I such that

$$\overline{\{u(x, t);\ (x, t) \in \Omega \times [0, \infty)\}} \subset I$$

and that

$$\text{for each } x \in \overline{\Omega}, \ f(x, u) \text{ is analytic for } u \in I. \tag{3.13}$$

As it is concerned with asymptotic behavior of the solution $u(t)$ lying in (3.12) only, we are allowed to cut off the values $f(x, u)$ for all u which are sufficiently large. For this reason, assume that

$$f(x, u) \equiv 0 \quad \text{for } (x, u) \in \overline{\Omega} \times \{u \in \mathbb{R}; \ |u| \geq \rho + 1\}, \tag{3.14}$$

$\rho > 0$ being a constant such that $[-\rho, \rho] \supset I$. Consequently, we see that $\mathcal{D}(f) = L_2(\Omega)$. Due to (3.14), $f(x, u)$ satisfies

$$|f(x, u) - f(x, v)| \leq L|u - v| \quad \text{for } x \in \overline{\Omega}; \ u, v \in \mathbb{R}, \tag{3.15}$$

with a constant $L > 0$. Thereby, $f(u)$ satisfies a global Lipschitz condition

$$\|f(u) - f(v)\|_{L_2} \leq L\|u - v\|_{L_2}, \qquad u, v \in L_2(\Omega). \tag{3.16}$$

3.2   Global Boundedness of Sobolev Norms for $u(t)$

This section is devoted to proving that the uniform boundedness of $\|u(t)\|_{L_\infty}$ for $0 \leq t < \infty$ (assumed in (3.12)) implies that, for suitable exponents $1 < p < \infty$ and integers $m \geq 1$, the global norm estimates $\|u(t)\|_{H_p^m} \leq C_{m,p}$ hold true with some constants $C_{m,p}$. These global estimates of Sobolev norms will play a crucial role in establishing the gradient inequality in Section 3.6. As the results are quite different for each dimension n, they will be discussed individually.

3.2.1   Case Where $n = 2$

When $n = 2$, we want to prove that, for the exponent p mentioned in (3.9), it holds that

$$\|u(t)\|_{H_p^1} \leq C_{1,p}, \qquad \tau_0 \leq \forall t < \infty, \tag{3.17}$$

$\tau_0 > 0$ being any fixed time. But this is verified immediately by applying the general theory of semilinear abstract evolution equations to (3.7).

In view of (3.16), it is possible to apply Theorem 1.4 with $\beta = \eta = 0$. We then obtain that the local norm estimate

$$\|Au(t)\|_{L_2} + \|u'(t)\|_{L_2} \leq C_{u(0)}/t, \qquad 0 < t \leq T_{u(0)},$$

is valid with a time $T_{u(0)} > 0$ and a constant $C_{u(0)} > 0$ which are both determined by the norm $\|u(0)\|_{L_2}$ of initial value $u(0)$ alone. Here, resetting the initial time to any other $\tau > 0$, we will regard $u(t)$ as a solution to (3.7) for $t \in [\tau, \infty)$ with initial value $u(\tau)$. It must then follow that

$$\|Au(t)\|_{L_2} + \|u'(t)\|_{L_2} \le C_{u(\tau)}/(t - \tau), \qquad \tau < t \le \tau + T_{u(\tau)},$$

$T_{u(\tau)} > 0$ and $C_{u(\tau)} > 0$ being determined by the norm $\|u(\tau)\|_{L_2}$ alone. As a matter of fact, because of (3.12), $C_{u(\tau)}$ and $T_{u(\tau)}$ are determined uniformly in the initial time $\tau\ (> 0)$; hence,

$$\|Au(t)\|_{L_2} \le R, \qquad \tau_0 \le \forall t < \infty, \tag{3.18}$$

with an arbitrarily fixed time $\tau_0 > 0$ and some constant R. Then, in view of (3.9), the estimate (3.17) is satisfied.

3.2.2   Case Where $n = 3$

In this case, our goal is to verify the global estimate

$$\|u(t)\|_{H^2} \le C_{2,2}, \qquad \tau_0 \le \forall t < \infty, \tag{3.19}$$

$\tau_0 > 0$ being any fixed time. But this is also verified immediately by applying the general results as in the case where $n = 2$. We just notice that the domain $\mathcal{D}(A)$ is characterized by (3.10).

3.2.3   Case Where $n = 4$

Our goal is to verify the estimate

$$\|u(t)\|_{H^3} \le C_{3,2}, \qquad \tau_1 \le \forall t < \infty, \tag{3.20}$$

$\tau_1 > 0$ being any fixed time.

In order to prove this, however, we have to formulate (3.1) (instead of (3.7)) as the abstract semilinear equation

$$u' + A_p u = f_p(u), \qquad 0 < t < \infty, \tag{3.21}$$

in the Banach space $L_p(\Omega)$ by introducing a fixed exponent $2 < p < \infty$. Here, A_p is a realization of $-\sum_{i,j=1}^{4} \frac{\partial}{\partial x_j}\left(a_{ij}(x)\frac{\partial}{\partial x_i}\right)$ in $L_p(\Omega)$ under the boundary conditions $u = 0$ on $\partial\Omega$. It is known that (3.4) together with (3.5) ($n = 4$) implies that $A_{\dot{p}}$ is a real sectorial operator of $L_p(\Omega)$ of angle $\omega_{A_p} < \frac{\pi}{2}$ having its domain such that

$$\mathcal{D}(A_p) \subset H_p^2(\Omega) \text{ (with continuous embedding).} \tag{3.22}$$

Meanwhile, $f_p(u)$ is an operator of $L_p(\Omega)$ defined by $f_p(u) = f(x, u)$. Due to (3.15), it is seen that that $\|f_p(u) - f_p(v)\|_{L_p} \leq L\|u - v\|_{L_p}$ for $u, v \in L_p(\Omega)$.

It is again possible to apply Theorem 1.4 with $\beta = \eta = 0$ to (3.21). In particular, we obtain that there exist a time $T_{u(0)} > 0$ and a constant $C_{u(0)} > 0$ determined by the norm $\|u(0)\|_{L_p}$ of initial value $u(0)$ alone for which the local norm estimate

$$\|A_p u(t)\|_{L_p} + \|u'(t)\|_{L_p} \leq C_{u(0)}/t, \qquad 0 < t \leq T_{u(0)},$$

is valid. Then, resetting the initial time to any other $\tau > 0$, we can regard $u(t)$ as a solution to (3.21) for $t \in [\tau, \infty)$ with initial vale $u(\tau)$. Then, it must follow that

$$\|A_p u(t)\|_{L_p} + \|u'(t)\|_{L_p} \leq C_{u(\tau)}/(t - \tau), \qquad \tau < t \leq \tau + T_{u(\tau)},$$

$T_{u(\tau)} > 0$ and $C_{u(\tau)} > 0$ being determined by the norm $\|u(\tau)\|_{L_p}$ alone. But, because of (3.12), $C_{u(\tau)}$ and $T_{u(\tau)}$ can be determined uniformly in initial time τ (> 0). In this way, $u(t)$ is verified to belong to

$$u \in \mathcal{C}([\tau_0, \infty); H_p^2(\Omega)) \cap \mathcal{C}^1([\tau_0, \infty); L_p(\Omega)) \tag{3.23}$$

and due to (3.22) satisfies the estimate

$$\|u(t)\|_{H_p^2} + \|u'(t)\|_{L_p} \leq C_{2,p}, \qquad \tau_0 \leq \forall t < \infty, \tag{3.24}$$

with an arbitrarily fixed time $\tau_0 > 0$.

We next prove the higher temporal regularity of $u(t)$.

Proposition 3.1 *With an arbitrarily fixed time $\tau_1 > \tau_0$, $u(t)$ satisfies*

$$u' \in \mathcal{C}([\tau_1, \infty); H_p^2(\Omega)) \cap \mathcal{C}^1([\tau_1, \infty); L_p(\Omega))$$

and satisfies the norm estimate

$$\|u'(t)\|_{H_p^2} + \|u''(t)\|_{L_p} \leq C'_{2,p}, \qquad \tau_1 \leq t < \infty, \tag{3.25}$$

$C'_{2,p} > 0$ *being some constant.*

Proof Before proving the differentiability of $u'(t)$ as an $L_p(\Omega)$-valued function, we notice by maximal regularity of solutions ([Yag10, (4.17)]) for initial data $u(\tau) \in$

$\mathcal{D}(A_p)$ that $u(t)$ and $u'(t)$ are locally, uniformly Hölder continuous for $\tau_0 \leq t < \infty$, that is,

$$\|u(t) - u(s)\|_{H_p^2} + \|u'(t) - u'(s)\|_{L_p} \leq L_0 |t - s|^{\sigma_0},$$

$$\tau_0 \leq \forall s, \ \forall t < \infty, \ |s - t| \leq 1, \qquad (3.26)$$

with some exponent $\sigma_0 > 0$ and some constant L_0.

We now introduce an auxiliary initial value problem

$$\begin{cases} v' + A_p v = \chi'(t), & \tau_0 < t < \infty, \\ v(\tau_0) = u'(\tau_0), \end{cases} \qquad (3.27)$$

in $L_p(\Omega)$ for an unknown function $v(t)$, where $\chi(t) \equiv f_p(u(t))$. By Corollary 1.2, $\chi(t)$ is continuously differentiable with the derivative $\chi'(t) = f_u(x, u(t))u'(t)$.

Since $H_p^2(\Omega) \subset \mathcal{C}(\overline{\Omega})$ with continuous embedding, it follows from (3.26) that $\chi'(t)$ is locally, uniformly Hölder continuous on $[\tau_0, \infty)$ as an $L_p(\Omega)$-valued function. Then, according to the result [Yag10, Theorem 3.5] for linear equations, there exists a unique solution $v(t)$ to the problem (3.27) which belongs to

$$v \in \mathcal{C}([\tau_0, \infty); L_p(\Omega)) \cap \mathcal{C}((\tau_0, \infty); \mathcal{D}(A_p)) \cap \mathcal{C}^1((\tau_0, \infty); L_p(\Omega)).$$

On the other hand, $v(t)$ is seen to coincide with the derivative $u'(t)$. In fact, for $\varepsilon > 0$, put $u_\varepsilon(t) = \varepsilon^{-1}[u(t + \varepsilon) - u(t)]$. Obviously, $u_\varepsilon(t)$ is a solution to the problem

$$\begin{cases} [u_\varepsilon]' + A_p u_\varepsilon = \chi_\varepsilon(t), & \tau_0 < t < \infty, \\ u_\varepsilon(\tau_0) = \varepsilon^{-1}[u(\tau_0 + \varepsilon) - u(\tau_0)], \end{cases}$$

in $L_p(\Omega)$, where $\chi_\varepsilon(t) = \varepsilon^{-1}[\chi(t + \varepsilon) - \chi(t)]$. Moreover, from (3.27),

$$\begin{cases} [u_\varepsilon - v]' + A_p[u_\varepsilon - v] = \chi_\varepsilon(t) - \chi'(t), & \tau_0 < t < \infty, \\ [u_\varepsilon - v](\tau_0) = \varepsilon^{-1}[u(\tau_0 + \varepsilon) - u(\tau_0)] - u'(\tau_0). \end{cases}$$

Then, [Yag10, (3.13)] provides the representation formula

$$[u_\varepsilon - v](t) = e^{-(t-\tau_0)A_p}[u_\varepsilon - v](\tau_0) + \int_{\tau_0}^t e^{-(t-s)A_p}[\chi_\varepsilon(s) - \chi'(s)]ds.$$

Here let $\varepsilon \searrow 0$. Since $[u_\varepsilon - v](\tau_0) \to 0$ in $L_p(\Omega)$ due to (3.23) and since $\chi_\varepsilon(t) - \chi'(t) \to 0$ in $L_p(\Omega)$ for any $t \geq \tau_0$ due to Corollary 1.2, we conclude that $u_\varepsilon(t) - v(t) \to 0$ in $L_p(\Omega)$, that is, $u'(t)$ coincides with $v(t)$ for every $t \geq \tau_0$.

Let us next verify (3.25). Thanks to [Yag10, Theorem 3.5] again, since $v(t) \equiv u'(t)$, we observe that

$$\|A_p u'(t)\|_{L_p} + \|u''(t)\|_{L_p} \leq C_{u'(\tau_0)}/(t - \tau_0), \qquad \tau_0 < t \leq \tau_0 + 1,$$

$C_{u'(\tau_0)}$ being determined by $\|u'(\tau_0)\|_{L_p}$ and the local Hölder norm of $\chi'(t)$ alone. But we can reset the initial time to any other $\tau > \tau_0$ and have

$$\|A_p u'(t)\|_{L_p} + \|u''(t)\|_{L_p} \leq C_{u'(\tau)}/(t - \tau), \qquad \tau < t \leq \tau + 1,$$

$C_{u'(\tau)}$ being determined by $\|u'(\tau)\|_{L_p}$ and the local Hölder norm of $\chi'(t)$. As (3.24) is already known, we see that the constant $C_{u'(\tau)}$ is uniform in the initial time τ ($>$ τ_0). As a consequence, in view of (3.22), we obtain the desired estimate (3.25). □

Finally, on the basis of (3.25), let us verify the global estimate (3.20). We use the relation $A_p u(t) = -u'(t) + f_p(u(t))$. By (3.25) we have $u'(t) \in H_p^2(\Omega)$ for $\tau_1 \leq t < \infty$. Meanwhile, it is easily verified by (3.6) ($n = 4$) that $f_p(u(t)) \in H_p^1(\Omega)$ for $t \geq \tau_0$ (cf. Theorem 1.3). Therefore, it follows for $t \geq \tau_1$ that $A_p u(t) \in H_p^1(\Omega)$. Then, owing to the assumption (3.5) ($n = 4$), the shift property [Yag10, Theorem 2.9] is available with $m = 1$ to a realization of $-\sum_{i,j=1}^4 \frac{\partial}{\partial x_j}\left(a_{ij}(x)\frac{\partial}{\partial x_i}\right)$ in $L_2(\Omega)$. Hence, it is concluded that $u(t) \in H^3(\Omega)$ for every $t \geq \tau_1$ together with (3.20).

3.2.4 Case Where $n \geq 5$

Let $n \geq 5$ be arbitrarily fixed. We choose an exponent p so that $\frac{n}{2} < p < \infty$ and work in the Sobolev spaces $H_p^{2m}(\Omega)$, where $m = 0, 1, 2, \ldots$. As is well known, $H_p^2(\Omega)$ is a Banach algebra, i.e., $\|uv\|_{H_p^2} \leq C\|u\|_{H_p^2}\|v\|_{H_p^2}$ for $u, v \in H_p^2(\Omega)$ (cf. [Yag10, P. 49, (3)]). For $k = 1, 2, 3, \ldots$, we denote the partial derivative $\frac{\partial^k f}{\partial u^k}(x, u)$ by $f^{(k)}(x, u)$. Due to (3.14), $f^{(k)}(x, u)$ also satisfies the same condition. Then, by Theorem 1.16, we see that

the operator $u \mapsto f^{(k)}(u)$ from $H_p^2(\Omega)$ into itself is continuously

Fréchet differentiable with the derivative $h \mapsto f^{(k+1)}(u)h$. (3.28)

Our goal is indeed to establish the estimate

$$\|u(t)\|_{H_p^{2(m+1)}} \leq C_{2(m+1),p}, \qquad \tau_m \leq \forall t < \infty, \tag{3.29}$$

for every $m = 0, 1, 2, \ldots$, τ_m being an arbitrarily fixed temporal sequence such that

$$\tau_{-1} = 0 < \tau_0 < \tau_1 < \tau_2 < \cdots < \tau_m < \tau_{m+1} < \cdots < 1.$$

We will first show the infinite temporal regularity of $u(t)$ by using (3.6) ($n \geq 5$) and then obtain the wanted estimate (3.29) by using (3.2) ($n \geq 5$) and (3.5) ($n \geq 5$).

3.2.5 Infinite Temporal Regularity

By induction on $m = 0, 1, 2, \ldots$, we will show that $u(t)$ is infinitely differentiable for t as an $H_p^2(\Omega)$-valued function.

Our induction consists of two assertions. The first one is that the m-th derivative $u^{(m)}(t)$ belongs to the function space

$$u^{(m)} \in \mathcal{C}([\tau_m, \infty); H_p^2(\Omega)) \cap \mathcal{C}^1([\tau_m, \infty); L_p(\Omega)) \tag{3.30}$$

together with the uniform estimates

$$\|u^{(m)}(t)\|_{H_p^2} + \|u^{(m+1)}(t)\|_{L_p} \leq C_m, \qquad \tau_m \leq \forall t < \infty, \tag{3.31}$$

with some constant C_m. In addition, $u^{(m)}(t)$ satisfies the locally uniform Hölder condition

$$\|u^{(m)}(t) - u^{(m)}(s)\|_{H_p^2} + \|u^{(m+1)}(t) - u^{(m+1)}(s)\|_{L_p}$$

$$\leq L_m |t - s|^{\sigma_m}, \qquad \tau_m \leq \forall s, \forall t < \infty, \; |s - t| \leq 1, \tag{3.32}$$

with some exponent $\sigma_m > 0$ and some constant L_m.

The second one is that $u^{(m)}$ is characterized as a solution to the Cauchy problem

$$\begin{cases} [u^{(m)}]' + A_p u^{(m)} = \chi^{(m)}(t), & \tau_{m-1} < t < \infty, \\ u^{(m)}(\tau_{m-1}) = u_m, \end{cases} \tag{3.33}$$

in $L_p(\Omega)$ with $u_m = u^{(m)}(\tau_{m-1})$. Here, A_p is a realization of $-\sum_{i,j=1}^n \frac{\partial}{\partial x_j}\left(a_{ij}\frac{\partial}{\partial x_i}\right)$ in $L_p(\Omega)$ under the boundary conditions $u = 0$ on $\partial\Omega$. Meanwhile, $\chi(t)$ denotes the function $f(u(t))$ whose m-th derivative is given by the formula

$$\chi^{(m)}(t) = \sum_{k=1}^{m} f^{(k)}(u(t)) \times$$

$$\left(\sum_{\substack{\sum_{\ell=1}^{m} p_{k\ell} = k, \\ \sum_{\ell=1}^{m} \ell p_{k\ell} = m}} C_{p_{k1}, p_{k2}, \dots, p_{km}} [u'(t)]^{p_{k1}} [u''(t)]^{p_{k2}} \cdots [u^{(m)}(t)]^{p_{km}} \right), \qquad (3.34)$$

where $0 \le p_{k1}, \, p_{k2}, \dots, \, p_{km} \le k$, by suitable coefficients $C_{p_{k1}, p_{k2}, \dots, p_{km}}$.

(I) *Case* $m = 0$. We utilize the L_p formulation of (3.1) (cf. (3.21)) instead of the
L_2 formulation, namely, $u(t)$ is regarded as a solution to the problem (3.33) of
the case $m = 0$. It is actually known that the operator A_p is a real sectorial
operator of $L_p(\Omega)$ of angle $\omega_{A_p} < \frac{\pi}{2}$ having its domain $\mathcal{D}(A_p)$ such that

$$\mathcal{D}(A_p) \subset H_p^2(\Omega) \quad \text{(with continuous embedding)}. \qquad (3.35)$$

Meanwhile, $u \mapsto f_p(u) \equiv f(x, u)$ is an operator from $L_p(\Omega)$ into itself and
(3.15) implies that $\| f_p(u) - f_p(v) \|_{L_p} \le L \| u - v \|_{L_p}$ for $u, \, v \in L_p(\Omega)$. We
can then apply Theorem 1.4 to (3.33) ($m = 0$) to see that there exist a constant
$C_{u(0)} > 0$ and a time $T_{u(0)} > 0$ determined by the norm $\| u(0) \|_{L_p}$ alone for
which the local norm estimate

$$\| A_p u(t) \|_{L_p} + \| u'(t) \|_{L_p} \le C_{u(0)}/t, \qquad 0 < t \le T_{u(0)},$$

is valid. Moreover, resetting the initial time to any other $\tau > 0$, we can regard
$u(t)$ as a solution for $t \in [\tau, \infty)$ with initial value $u(\tau)$. Then, it follows that

$$\| A_p u(t) \|_{L_p} + \| u'(t) \|_{L_p} \le C_{u(\tau)}/(t - \tau), \qquad \tau < t \le \tau + T_{u(\tau)},$$

$C_{u(\tau)} > 0$ and $T_{u(\tau)} > 0$ being determined by the norm $\| u(\tau) \|_{L_p}$ alone. But,
because of (3.12), $C_{u(\tau)}$ and $T_{u(\tau)}$ can be determined uniformly in initial time
$\tau \, (> 0)$. In this way, $u(t)$ is verified to belong to the space of (3.30) and to
satisfy the estimate

$$\| A_p u(t) \|_{L_p} + \| u'(t) \|_{L_p} \le C_0, \qquad \tau_0 \le \forall t < \infty, \qquad (3.36)$$

with some constant C_0, which means that (3.31) holds true.

In order to verify (3.32), we appeal to the maximal regularity [Yag10, (4.17)]
for the solutions to semilinear equations. Since $u(\tau) \in \mathcal{D}(A_p)$ for any initial
time $\tau \ge \tau_0$ and since (3.36) holds for $u(\tau)$, $A_p u(t)$ and $u'(t)$ are seen to be

locally, uniformly Hölder continuous for $\tau_0 \leq t < \infty$ in such a way that

$$\|A_p[u(t) - u(s)]\|_{L_p} + \|u'(t) - u'(s)\|_{L_p} \leq L_0|t - s|^{\sigma_0},$$

$$\tau_0 \leq \forall s, \ \forall t < \infty, \ |s - t| \leq 1,$$

with some exponent $\sigma_0 > 0$ and some constant L_0.

When $m = 0$, (3.34) is trivial.   □

(II) *Case $m + 1$.* Assume the assertions of induction are proved for all integers k such that $0 \leq k \leq m$.

We begin by verifying the differentiability of $\chi^{(m)}(t)$.

Lemma 3.1 *As an $L_p(\Omega)$-valued function, $\chi^{(m)}(t)$ is differentiable.*

Proof of Lemma By Corollary 1.2, $f^{(k)}(u(t))$ is differentiable as an $L_p(\Omega)$-valued function. The condition (3.30) implies that, for $1 \leq k \leq m - 1$, $u^{(k)}(t)$ is differentiable as an $H_p^2(\Omega)$-valued function, $H_p^2(\Omega)$ being embedded in $\mathcal{C}(\overline{\Omega})$; and, $u^{(m)}(t)$ is continuous as an $H_p^2(\Omega)$-valued function and is differentiable as an $L_p(\Omega)$-valued function. By these we can verify that $\chi^{(m)}(t)$ is differentiable as an $L_p(\Omega)$-valued function. The derivative of the k-th term in the right hand side of (3.34) becomes

$$\{f^{(k)}(u(t))[u'(t)]^{p_{k1}}[u''(t)]^{p_{k2}} \cdots [u^{(m)}(t)]^{p_{km}}\}'$$

$$= f^{(k+1)}(u)[u']^{p_{k1}+1}[u'']^{p_{k2}} \cdots [u^{(m)}]^{p_{km}}$$

$$+ \sum_{\ell=1}^{m} p_{k\ell} f^{(k)}(u)[u']^{p_{k1}} \cdots [u^{(\ell-1)}]^{p_{k(\ell-1)}}[u^{(\ell+1)}]^{p_{k\ell}+1} \cdots [u^{(m)}]^{p_{km}}.$$

Therefore, $\chi^{(m+1)}(t)$ is written as

$$\chi^{(m+1)}(t) = \sum_{k=1}^{m+1} f^{(k)}(u(t)) \times$$

$$\left(\sum_{\substack{\sum_{\ell=1}^{m+1} p_{k\ell} = k, \\ \sum_{\ell=1}^{m+1} \ell p_{k\ell} = m+1}} C_{p_{k1}, p_{k2}, \ldots, p_{k(m+1)}} [u'(t)]^{p_{k1}}[u''(t)]^{p_{k2}} \cdots [u^{(m+1)}(t)]^{p_{k(m+1)}} \right).$$

$$(3.37)$$

This equality shows also that the formula (3.34) is valid for $m + 1$, too.   □

Furthermore, we notice that $\chi^{(m+1)}(t)$ is locally, uniformly Hölder continuous on $[\tau_m, \infty)$. In fact, in (3.37), the integer k varies from 1 to $m + 1$. If $k = 1$, then $p_{11} = p_{12} = \cdots = p_{1m} = 0$ and $p_{1(m+1)} = 1$; therefore, the corresponding term in the right hand side is just $f'(u(t))u^{(m+1)}(t)$. Since the operator $u \mapsto f'(u)$ satisfies (3.28) and since $u^{(m)}(t)$ satisfies (3.32), $f'(u(t))u^{(m+1)}(t)$ is observed to be locally, uniformly Hölder continuous on $[\tau_m, \infty)$. Meanwhile, if $k \geq 2$, then $p_{k(m+1)} = 0$; therefore, the corresponding terms are all of the form $f^{(k)}(u(t))[u'(t)]^{p_{k1}}[u''(t)]^{p_{k2}} \cdots [u^{(m)}(t)]^{p_{km}}$. Then, due to (3.32), these functions are also locally, uniformly Hölder continuous.

The assumptions (3.31) and (3.32) of induction thus imply that $\chi^{(m+1)}(t)$ is locally, uniformly Hölder continuous on $[\tau_m, \infty)$ as an $L_p(\Omega)$-valued function.

In order to verify the differentiability of $u^{(m+1)}(t)$, we use an auxiliary problem

$$
\begin{cases}
v' + A_p v = \chi^{(m+1)}(t), & \tau_m < t < \infty, \\
v(\tau_m) = u^{(m+1)}(\tau_m),
\end{cases}
\tag{3.38}
$$

in $L_p(\Omega)$ for an unknown function $v(t)$. As $\chi^{(m+1)}(t)$ is a Hölder continuous function on $[\tau_m, \infty)$, the result for linear equations is available as before. There exists a unique solution to (3.38) in the function space

$$
v \in \mathcal{C}([\tau_m, \infty); L_p(\Omega)) \cap \mathcal{C}((\tau_m, \infty); \mathcal{D}(A_p)) \cap \mathcal{C}^1((\tau_m, \infty); L_p(\Omega)).
$$

By the same arguments as in the proof of Proposition 3.1, $v(t)$ is observed to be the derivative $u^{(m+1)}(t)$, that is, $u^{(m+1)}(t)$ is differentiable with the derivative $u^{(m+2)}(t) \equiv v'(t)$ in (τ_m, ∞). Moreover, as before, $u^{(m+1)}(t)$ satisfies the local norm estimate

$$
\|A_p u^{(m+1)}(t)\|_{L_p} + \|u^{(m+2)}(t)\|_{L_p} \leq C_{u_{m+1}, \chi^{(m+1)}}/(t - \tau_m), \quad \tau_m < t \leq \tau_m + 1,
$$

$C_{u_{m+1}, \chi^{(m+1)}}$ being determined by the norm $\|u_{m+1}\|_{L_p} = \|u^{(m+1)}(\tau_m)\|_{L_p}$ and by the Hölder norm of $\chi^{(m+1)}(t)$ alone. Resetting the initial time to any other $\tau > \tau_m$ and regarding $u^{(m+1)}(t)$ as a solution for $t \in [\tau, \infty)$ with initial value $u^{(m+1)}(\tau)$, we can verify the estimate (3.31) for $u^{(m+1)}(t)$ with any $\tau_{m+1} > \tau_m$. The Hölder estimate (3.32) for $u^{(m+1)}(t)$ is also verified by the same arguments as above by utilizing the maximal regularity for the solution $v(t)$ of the linear evolution equation (3.38).

3.2.6 Infinite Spatial Regularity

On the basis of established temporal regularity in $L_p(\Omega)$, we now show spatial infinite regularity for $u(t)$.

We use again an induction on $m = 0, 1, 2, \ldots$. The assertion of induction for m is that the derivatives $u^{(k)}(t)$ up to m belong to

$$u^{(k)} \in \mathcal{C}([\tau_m, \infty); H_p^{2(m+1-k)}(\Omega)), \qquad 0 \leq \forall k \leq m, \tag{3.39}$$

respectively, and satisfy the estimate

$$\sum_{k=0}^{m} \|u^{(k)}(t)\|_{H_p^{2(m+1-k)}} \leq D_m, \qquad \tau_m \leq \forall t < \infty, \tag{3.40}$$

with some constant D_m.

(I) *Case* $m = 0$. The assertions are already proved.
(II) *Case* $m + 1$. Assume that (3.39) and (3.40) are proved for an integer $m \geq 0$. According to (3.33), we have $A_p u^{(m)}(t) = -u^{(m+1)}(t) + \chi^{(m)}(t)$. Here, due to (3.30), $u^{(m+1)}(t) \in H_p^2(\Omega)$ for $\tau_{m+1} \leq t < \infty$. Meanwhile, we have the following lemma.

Lemma 3.2 *The function* $\chi^{(m)}(t)$ *takes its values in* $H_p^2(\Omega)$ *for* $\tau_m \leq t < \infty$.

Proof (Proof of lemma) By (3.34), $\chi^{(m)}(t)$ is a sum of functions of the product $f^{(k)}(u)[u']^{p_{k1}}[u'']^{p_{k2}} \cdots [u^{(m)}]^{p_{km}}$. But we already know (3.30) and (3.31) for any integer m and the fact that $H_p^2(\Omega)$ is a Banach algebra. Then, the assertion of the lemma is obvious. $\square$

Now, since $A_p u^{(m)}(t) \in H_p^2(\Omega)$, the higher shift property [Tri78, Theorem 5.4.1] applied for A_p yields that $u^{(m)}(t)$ certainly takes its values in $H_p^4(\Omega)$ for $\tau_{m+1} \leq t < \infty$.

In the next step, we use the relation $A_p u^{(m-1)} = -u^{(m)}(t) + \chi^{(m-1)}(t)$. As just seen, we have $u^{(m)}(t) \in H_p^4(\Omega)$ for $\tau_{m+1} \leq t < \infty$. Meanwhile, we observe that $\chi^{(m-1)}(t)$ also takes its values in $H_p^4(\Omega)$, for the operator $u \mapsto f^{(k)}(u)$ maps $H_p^4(\Omega)$ into itself and for $H_p^4(\Omega)$ is a Banach algebra. Then, since $A_p u^{(m-1)}(t) \in H_p^4(\Omega)$, the higher shift property provides that $u^{(m-1)}(t) \in H_p^6(\Omega)$ for $\tau_{m+1} \leq t < \infty$.

We can repeat these arguments step by step until arriving at the spatial regularity $u(t) \in H_p^{2(m+2)}(\Omega)$ for $\tau_{m+1} \leq t < \infty$.

By the same procedure, (3.40) can also be proved for $m + 1$.

In particular, (3.40) provides the important estimate

$$\|u(t)\|_{H^{2(m+1)}} \leq D_m, \qquad 1 \leq t < \infty, \quad m = 0, 1, 2, \ldots. \tag{3.41}$$

3.3 Some Other Properties of $u(t)$

Let us observe properties of the solution $u(t)$. Their proof is independent of the dimension n.

3.3.1 Lyapunov Function

Take the inner product between $u'(t)$ and Eq. (3.7) in $L_2(\Omega)$. Then,

$$\|u'(t)\|_{L_2}^2 + (Au(t), u'(t))_{L_2} = (f(u(t)), u'(t))_{L_2}.$$

Here, since

$$\frac{\|A^{\frac{1}{2}}u(t+\Delta t)\|_{L_2}^2 - \|A^{\frac{1}{2}}u(t)\|_{L_2}^2}{\Delta t} = \left(A[u(t+\Delta t) + u(t)], \frac{u(t+\Delta t) - u(t)}{\Delta t} \right)_{L_2},$$

it follows by (3.12) that

$$\frac{d}{dt}\|A^{\frac{1}{2}}u(t)\|_{L_2}^2 = 2(Au(t), u'(t))_{L_2}, \qquad 0 \le t < \infty.$$

Meanwhile, put $F(x, u) = \int_{u_*}^{u} f(x, v)dv$ for $-\infty < u < \infty$, u_* being the central point of I. Then, due to (3.14), we have $F(x, u) \equiv F(x, -(\rho+1))$ for $u \le -(\rho+1)$ and $F(x, u) \equiv F(x, \rho+1)$ for $u \ge \rho+1$; therefore, $F(x, u)$ belongs to $\mathcal{C}_{\infty}^{0,1}(\overline{\Omega}, \mathbb{R})$. Corollary 1.3 is then available to the integral $\int_{\Omega} F(x, u(t))dx$ to observe that

$$\frac{d}{dt}\int_{\Omega} F(x, u(t))dx = \int_{\Omega} f(x, u(t))u'(t)dx = (f(u(t)), u'(t))_{L_2}, \qquad 0 \le t < \infty.$$

In this way, we obtain that

$$\frac{d}{dt}\Phi(u(t)) = -\|u'(t)\|_{L_2}^2, \qquad 0 \le t < \infty, \tag{3.42}$$

where, on account of (3.8) and (3.11),

$$\Phi(u) = \int_{\Omega} \left(\frac{1}{2}\sum_{i,j=1}^{n} a_{ij}(x)\frac{\partial u}{\partial x_i}\frac{\partial u}{\partial x_j} - F(x, u) \right) dx, \qquad u \in \mathring{H}^1(\Omega). \tag{3.43}$$

Thereby, the value $\Phi(u(t))$ decreases monotonously as $t \nearrow \infty$. But it is clear that

$$\lim_{t \to \infty} \Phi(u(t)) = \inf_{0 \le t < \infty} \Phi(u(t)) > -\infty. \tag{3.44}$$

Furthermore, (3.42) implies that

if $\frac{d}{dt}\Phi(u(t)) = 0$ at $t = \bar{t}$, then $Au(\bar{t}) = f(u(\bar{t}))$,

i.e., $u(\bar{t})$ is a stationary solution of (3.7). $\tag{3.45}$

3.3.2 ω-Limit Set

Consider $u(t)$'s ω-limit set

$$\omega(u) = \{\bar{u} \in L_2(\Omega); \ \exists t_m \nearrow \infty \text{ such that } u(t_m) \to \bar{u} \text{ in } L_2(\Omega)\}.$$

Since $H^1(\Omega)$ is compactly embedded in $L_2(\Omega)$ (cf. [Yag10, Theorem 1.38]) and since $u(t)$ satisfies the global norm estimate (3.17), (3.19), (3.20) and (3.29) for $n = 2, 3, 4$ and $n \geq 5$, respectively, it is concluded that $\omega(u) \neq \emptyset$.

As seen, any ω-limit $\bar{u}$ given as in (2.11) belongs to $\mathcal{D}(A)$ and there exists a temporal sequence $t_m \nearrow \infty$ such that $Au(t_m) \to A\bar{u}$ weakly and $u(t_m) \to \bar{u}$ strongly in $L_2(\Omega)$ respectively.

Proposition 3.2 *There exists an ω-limit $\bar{u} \in \omega(u)$ which is a stationary solution of* (3.7).

Proof Integrating the equality (3.42) in the half line $[0, \infty)$, we have

$$\int_0^{\infty} \|u'(t)\|_{L_2}^2 dt = \Phi(u(0)) - \lim_{t \to \infty} \Phi(u(t)) < \infty, \tag{3.46}$$

due to (3.44). Then, for any number $\varepsilon > 0$ and any $T > 0$, there must exist a time $\tau > T$ such that $\|u'(\tau)\|_{L_2}^2 < \varepsilon$; as a consequence, there exists a temporal sequence $t_m \nearrow \infty$ for which it holds that $u'(t_m) \to 0$ in $L_2(\Omega)$. Then, there exist a subsequence $t_{m'}$ of t_m and an ω-limit $\bar{u}$ such that $Au(t_{m'}) \to A\bar{u}$ weakly and $u(t_{m'}) \to \bar{u}$ strongly. On account of (3.16), $u(t_{m'}) \to \bar{u}$ in $L_2(\Omega)$ implies $f(u(t_{m'})) \to f(\bar{u})$ in $L_2(\Omega)$. Therefore, letting $m' \to \infty$ in the equation $u'(t_{m'}) + Au(t_{m'}) = f(u(t_{m'}))$, we obtain that $A\bar{u} = f(\bar{u})$. □

3.4 Formulation

Let A be the realization of $-\sum_{i,j=1}^n \frac{\partial}{\partial x_j} \left(a_{ij}(x) \frac{\partial}{\partial x_i}\right)$ in $L_2(\Omega)$ under the boundary conditions $u = 0$ on $\partial\Omega$ introduced above. By assumption, our solution $u(t)$ belongs to $\mathcal{C}([0, \infty); \mathcal{D}(A)) \cap \mathcal{C}^1([0, \infty); L_2(\Omega))$. Meanwhile, the Lyapunov function $\Phi(u)$ given by (3.43) is defined on $\overset{\circ}{H}{}^1(\Omega) = \mathcal{D}(A^{\frac{1}{2}})$ (remember (3.11)). Taking account of these situations, we are naturally led to set the triplet $Z \subset X \subset Z^*$ as

$$Z = \mathcal{D}(A), \quad X = \mathcal{D}(A^{\frac{1}{2}}), \quad Z^* = \mathcal{D}(A^0) = L_2(\Omega). \tag{3.47}$$

Obviously, Z (resp. X) is a Hilbert space equipped with the inner product $(\cdot, \cdot)_Z = (A\cdot, A\cdot)_{L_2}$ (resp. $(\cdot, \cdot)_X = (A^{\frac{1}{2}}\cdot, A^{\frac{1}{2}}\cdot)_{L_2}$). So, the duality product between Z and

Z^* is observed to become

$$\langle u, f \rangle_{Z \times Z^*} = (Au, f)_{L_2}, \qquad \forall (u, f) \in Z \times Z^*$$

$$\langle u, v \rangle_{Z \times Z^*} = (A^{\frac{1}{2}}u, A^{\frac{1}{2}}v)_{L_2}, \qquad \forall (u, v) \in Z \times X.$$

The solution $u(t)$ satisfies, in addition to (2.3), the global norm estimate (2.4), because of (3.18), (3.19), (3.20) and (3.29) for $n = 2, 3, 4$ and $n \geq 5$, respectively.

The Fréchet differentiability of $\Phi : X \rightarrow \mathbb{R}$ is verified by the following proposition.

Proposition 3.3 *The function $\Phi(u)$ is continuously differentiable in X in the sense of (2.5)–(2.6) with the derivative*

$$\dot{\Phi}(u) = u - A^{-1} f(x, u), \qquad u \in X. \tag{3.48}$$

Furthermore, if $u \in Z$, then $\dot{\Phi}(u) \in Z$ and the mapping $u \mapsto \dot{\Phi}(u)$ is continuous from Z into itself.

Proof On account of (3.8), we can write

$$\Phi(u) = \frac{1}{2}(A^{\frac{1}{2}}u, A^{\frac{1}{2}}u)_{L_2} - (F(x, u), 1)_{L_2}, \qquad u \in \overset{\circ}{H}{}^1(\Omega).$$

Therefore, for $u, h \in \overset{\circ}{H}{}^1(\Omega)$,

$$\Phi(u + h) - \Phi(u) = (A^{\frac{1}{2}}u, A^{\frac{1}{2}}h)_{L_2} + \frac{1}{2}(A^{\frac{1}{2}}h, A^{\frac{1}{2}}h)_{L_2}$$
$$- (F(x, u + h) - F(x, u), 1)_{L_2}.$$

Here, we have $(A^{\frac{1}{2}}u, A^{\frac{1}{2}}h)_{L_2} = (u, h)_X$ as noticed above. Meanwhile, since $\overset{\circ}{H}{}^1(\Omega) \subset L_{q_n}(\Omega)$, where q_2 is any finite number and $q_n = \frac{2n}{n-2}$ for $n \geq 3$, Corollary 1.1 is available with $q > 2$ to obtain that

$$(F(x, u + h) - F(u), 1)_{L_2} = (f(x, u), h)_{L_2} + o(\|h\|_{L_{q_n}})$$
$$= (A^{-1}f(x, u), h)_X + o(\|h\|_X).$$

Hence,

$$\Phi(u + h) - \Phi(u) - (u - A^{-1}f(u), h)_X = o(\|h\|_X), \qquad u, h \in X,$$

which shows that $\Phi(u)$ is Fréchet differentiable and its derivative is given by (3.48).

Continuity of $u \mapsto \dot{\Phi}(u)$ in X follows from that of the mapping $u \mapsto f(x, u)$ from $L_q(\Omega)$ into $L_2(\Omega)$, which was verified in Theorem 1.14.

Finally, $\dot{\Phi}(u) \in Z$ for $u \in Z$ and continuity of $u \mapsto \dot{\Phi}(u)$ in Z is obvious. $\square$

As for the ω-limit lying in the set of (2.11), we choose the $\bar{u}$ which is a stationary solution of (3.7), whose existence was already shown by Proposition 3.2.

In the subsequent sections, we shall show that the structural assumptions announced in Section 2.2 are all fulfilled by this $\bar{u} \in \omega(u)$.

3.5 Verification of Structural Assumptions

In this section, let us verify the Critical Condition, Lyapunov Function and Angle Condition.

(I) *Critical Condition.* The ω-limit $\bar{u}$ has been chosen so that $\bar{u}$ is a stationary solution to (3.7). The formula (3.48) then implies that $\dot{\Phi}(\bar{u}) = 0$ is satisfied.

(II) *Lyapunov Function.* By (3.42) we already know that $\frac{d}{dt}\Phi(u(t)) \leq 0$ for any $0 \leq t < \infty$. According to (3.45), if $\frac{d}{dt}\Phi(u(t)) = 0$ at some time $t = \bar{t}$, then $u(\bar{t})$ is a stationary solution of (3.7) and $u(t) = u(\bar{t})$ for all $t \geq \bar{t}$. Thereby, convergence of the solution is automatically valid. For this reason, it suffices to argue under (2.12).

(III) *Angle Condition.* By (3.47) and (3.48), we can write

$$-\langle \dot{\Phi}(u(t)), u'(t) \rangle_{Z \times Z^*} = -(A\dot{\Phi}(u(t)), u'(t))_{L_2}$$
$$= \| - Au(t) + f(u(t)) \|_{L_2}^2 = \| u'(t) \|_{L_2}^2$$
$$= \| \dot{\Phi}(u(t)) \|_Z \| u'(t) \|_{Z^*},$$

which shows that (2.13) is fulfilled with $\delta = 1$.

Hence, if we verify the *Gradient Inequality*, then Theorem 2.1 can provide the asymptotic convergence of $u(t)$. But, as its verification requires much more essential considerations on $\Phi(u)$, the proof will be described in the next section.

3.6 Gradient Inequality

Let $\Phi(u)$ be the function given by (3.43) on the space $X = \overset{\circ}{H}^1(\Omega)$. In order to verify its Gradient Inequality in a neighborhood of the ω-limit $\bar{u} \in \omega(u)$, let us apply the general results reviewed in Section 2.3.

We begin by showing the following proposition.

Proposition 3.4 *The mapping $\dot{\Phi} : X \to X$ is continuously Fréchet differentiable. For any $u \in X$, the derivative is given by*

$$[\dot{\Phi}]'(u)h = h - A^{-1}f_u(x, u)h, \qquad h \in X. \tag{3.49}$$

Proof As seen from Proposition 3.3, $\Phi(u)$ is continuously differentiable and its derivative $\dot{\Phi}(u)$ is given by (3.48). Thereby, it suffices to discuss the differentiability for $u \mapsto A^{-1} f(x, u)$.

However, this mapping can be considered as a composition of mappings $u \mapsto u$ from $\overset{\circ}{H}{}^1(\Omega)$ into $L_{q_n}(\Omega)$, where q_2 is any finite number and $q_n = \frac{2n}{n-2}$ when $n \geq 3$, $u \mapsto f(x, u)$ from $L_{q_n}(\Omega)$ into $L_2(\Omega)$, and $u \mapsto A^{-1}u$ from $L_2(\Omega)$ into $\overset{\circ}{H}{}^1(\Omega)$. Except $u \mapsto f(x, u)$, the other two mappings are linear bounded operators. Therefore, the argument is essentially reduced to prove the differentiability of the mapping $u \mapsto f(x, u)$ but is directly observed by Theorem 1.14. $\square$

3.6.1   Verification of (2.18)

Let us see that $L = [\dot{\Phi}]'(\overline{u})$ is a Fredholm operator of X. Since $L = I - A^{-1} f_u(x, \overline{u})$ due to (3.49), I being the identity of X, Theorem 1.19 provides the result only if the following proposition is verified.

Proposition 3.5 *The operator* $h \mapsto A^{-1} f_u(x, \overline{u})h$ *is a compact operator from X into itself.*

Proof Since $h \mapsto f_u(x, \overline{u})h$ is a bounded operator from $\overset{\circ}{H}{}^1(\Omega)$ into $L_2(\Omega)$, it suffices to prove that A^{-1} is a compact operator from $L_2(\Omega)$ into $\overset{\circ}{H}{}^1(\Omega)$.

Consider any bounded sequence h_m of $L_2(\Omega)$. Then, $u_m = A^{-1}h_m$ is a bounded sequence of $\mathcal{D}(A)$. Since the embedding $\mathcal{D}(A) \subset L_2(\Omega)$ is compact, we can extract a subsequence $u_{m'}$ which is convergent in $L_2(\Omega)$. Here we employ the moment inequality to A (see [Yag10, (2.119)]) which gives that $\|A^{\frac{1}{2}}u\|_{L_2} \leq C\|Au\|_{L_2}^{\frac{1}{2}}\|u\|_{L_2}^{\frac{1}{2}}$ for all $u \in \mathcal{D}(A)$. It then follows that $u_{m'}$ in convergent even in $\mathcal{D}(A^{\frac{1}{2}})$. In view of (3.11), we conclude the desired compactness of A^{-1}. $\square$

3.6.2   Space Y

Let us now set the key space Y. Taking account of all the properties required for Y, that is, (2.19), (2.20), (2.21), (2.22) and (2.35), it is natural to set it as

$$Y = \overset{\circ}{H}{}^1(\Omega) \cap \mathcal{C}(\overline{\Omega}) \tag{3.50}$$

with the norm $\| \cdot \|_Y = \| \cdot \|_{H^1} + \| \cdot \|_{\mathcal{C}}$. Clearly, Y is a dense subspace of X with continuous embedding, i.e., (2.19) is fulfilled.

To verify (2.20) we notice that $\bar{u} = A^{-1} f(x, \bar{u})$ and that $f(x, \bar{u}) \in L_\infty(\Omega)$. Therefore, it suffices to verify that, for any dimension n,

$$A^{-1} \text{ maps } L_\infty(\Omega) \text{ into } \mathcal{C}(\overline{\Omega}). \tag{3.51}$$

Indeed, when $n = 2$, (3.9) implies $A^{-1} : L_2(\Omega) \to H_p^1(\Omega) \subset \mathcal{C}(\overline{\Omega})$ for some $p > 2$. Similarly, when $n = 3$, (3.10) implies that $A^{-1} : L_2(\Omega) \to H^2(\Omega) \subset \mathcal{C}(\overline{\Omega})$. Meanwhile, when $n \geq 4$, (3.22) and (3.35) imply that $A_p^{-1} : L_p(\Omega) \to H_p^2(\Omega)$ for some $p > \frac{n}{2}$. Since $A^{-1} = A_p^{-1}$ on $L_p(\Omega)$, it follows that $A^{-1}(L_\infty(\Omega)) \subset H_p^2(\Omega) \subset \mathcal{C}(\overline{\Omega})$. In this way, (2.20) is verified.

The condition (2.21) is also verified by similar arguments. Let $h \in L^{-1}(Y)$ for $h \in X$; then, since $h = Lh + A^{-1} f_u(x, \bar{u}) h$ and $Lh \in Y$, it clearly suffices to prove that $A^{-1} f_u(x, \bar{u}) h$ lies in $\mathcal{C}(\overline{\Omega})$. However, for the case when $n = 2$ or 3, this immediately follows from the fact that A^{-1} maps $L_2(\Omega)$ into $\mathcal{C}(\overline{\Omega})$. When $n = 4$, $h \in Y \subset \overset{\circ}{H}{}^1(\Omega)$ implies that $h \in L_p(\Omega)$ with some $p > 2$; then, it follows by (3.22) that $A^{-1} f_u(x, \bar{u}) h \in H_p^2(\Omega) \subset \mathcal{C}(\overline{\Omega})$. For the case when $n \geq 5$, we have to use iteration. For $2 \leq q < \frac{n}{2}$, assume that $h \in L_q(\Omega)$ is seen. Then, by (3.35), we have $A^{-1} f_u(x, \bar{u}) h \in H_q^2(\Omega) \subset L_{q'}(\Omega)$, where $\frac{1}{q'} = \frac{1}{q} - \frac{2}{n}$. Repeating this procedure, we obtain for finite times that $h \in L_p(\Omega)$ for some $p > \frac{n}{2}$. Then, by (3.35) again, $A^{-1} f_u(x, u) h \in H_p^2(\Omega) \subset \mathcal{C}(\overline{\Omega})$ is concluded.

The conditions (2.22) and (2.23) are verified by the following proposition.

Proposition 3.6 *The operator $u \mapsto \dot{\Phi}(u)$ maps Y into itself and is continuously Fréchet differentiable in Y with the derivative $[\dot{\Phi}]'(u) \in \mathcal{L}((Y)$ such that*

$$[\dot{\Phi}]'(u) h = h - A^{-1} f_u(x, u) h, \qquad h \in Y. \tag{3.52}$$

Proof Remember that Y was defined by (3.50). According to Proposition 3.4, the mapping $\dot{\Phi} : \overset{\circ}{H}{}^1(\Omega) \to \overset{\circ}{H}{}^1(\Omega)$ is continuously Fréchet differentiable. So, it suffices to verify that $\dot{\Phi}$ maps $\overset{\circ}{H}{}^1(\Omega) \cap \mathcal{C}(\overline{\Omega})$ into $\mathcal{C}(\overline{\Omega})$ and is continuously Fréchet differentiable in $\overset{\circ}{H}{}^1(\Omega) \cap \mathcal{C}(\overline{\Omega})$.

But, in view of (3.48), the first assertion follows immediately from (3.51). Meanwhile, the second one is directly verified by Theorem 1.15. In accordance with (3.49), the derivative $[\dot{\Phi}]'(u)$ is of course given by (3.52).   □

We have thus verified, except (2.35), all other structural assumptions in Section 2.3, i.e., (2.5)–(2.6), (2.16), (2.17), (2.18), (2.19), (2.20), (2.21), (2.22) and (2.23).

3.6.3 Verification of (2.35)

It now remains to verify (2.35). For this purpose, the straightforward way may be to appeal to the theory of analytic operators (see [Zei88, Definition 8.8]) and to argue that, as the operator $G(u_0, u_1)$ utilized in the proof of Proposition 2.2 is an analytic operator, its implicit function $g(u_0)$ of $G(u_0, u_1) = 0$ must be analytic, too. We argued in such a way in the paper [GMY]. But, the implicit function theorem of analytic version is not elementary and in order to apply it, heavy calculations are required. So, we will here prefer to use the method of complexification, that is, we extend everything used for defining the critical manifold S to the complex form and consider a complex critical manifold $S_\mathbb{C}$ whose real part is just S. As $S_\mathbb{C}$ becomes a differentiable complex manifold, the characterization of analytic functions of several complex variables may provide the analyticity to be verified for the function $\phi(\xi)$ in (2.35).

Let us recall that the critical manifold of $\Phi(u)$ was defined by

$$S = \{u \in Y; \ (I - P)\dot{\Phi}(u) = 0\} \tag{3.53}$$

(see (2.28)). Here, P is an orthogonal projection from X onto $\mathcal{K}(L)$ which is a finite-dimensional subspace of X due to (2.18), and induces an orthogonal decomposition $X = \mathcal{K}(L) + L(X)$. Due to (2.21), $\mathcal{K}(L)$ is included in Y. On the other hand, due to (2.22), L is a mapping from Y into itself. The decomposition (2.27) then provides that the projection P induces a topological decomposition of the Banach space Y, too, into the form

$$Y = \mathcal{K}(L) + L(Y), \tag{3.54}$$

$$P(Y) = \mathcal{K}(L) \quad \text{and} \quad (I - P)(Y) = L(Y). \tag{3.55}$$

Moreover, L is an isomorphism from $L(Y)$ onto itself. As proved by Proposition 2.2, these facts yield that, in a neighborhood of $\overline{u}$ on S, S is a $\mathbb{C}^1$ manifold having the same dimension as $\mathcal{K}(L)$. More precisely, there exists an open neighborhood $U = U_0 \times U_1$ of $\overline{u}$ in Y, where U_0 (resp. U_1) is an open neighborhood of $P\overline{u}$ (resp. $(I - P)\overline{u}$) in $\mathcal{K}(L)$ (resp. $L(Y)$), such that S is represented in U by

$$S \cap U = \{(u_0, g(u_0)); u_0 \in U_0, \ g : U_0 \to U_1\},$$

g being a $\mathbb{C}^1$ mapping from U_0 into U_1 satisfying $g(P\overline{u}) = (I - P)\overline{u}$.

Let $v_1, v_2, \ldots, v_N$ be a basis of $\mathcal{K}(L)$, where $N = \dim \mathcal{K}(L)$, and identify $\mathcal{K}(L)$ with $\mathbb{R}^N$ by the correspondence

$$u_0 = \sum_{k=1}^{N} \xi_k v_k \in \mathcal{K}(L) \quad \longleftrightarrow \quad \xi = (\xi_1, \xi_2, \ldots, \xi_N) \in \mathbb{R}^N.$$

Let $P\overline{u} \leftrightarrow \overline{\xi}$ and let U_0 correspond to an open neighborhood Ω of $\overline{\xi}$ in $\mathbb{R}^N$. Our goal is then to verify that

the function $\xi \in \Omega \mapsto \phi(\xi) \equiv \Phi\left(\sum_{k=1}^{N} \xi_k v_k + g(\sum_{k=1}^{N} \xi_k v_k)\right)$

is analytic in a neighborhood of $\overline{\xi}$. $\tag{3.56}$

We will begin by setting the complex spaces $X_{\mathbb{C}}$ and $Y_{\mathbb{C}}$ which are given in view of (3.47) and (3.50) by

$$X_{\mathbb{C}} = \mathring{H}^1(\Omega; \mathbb{C}) \quad \text{and} \quad Y_{\mathbb{C}} = \mathring{H}^1(\Omega; \mathbb{C}) \cap \mathcal{C}(\overline{\Omega}; \mathbb{C}) \qquad (3.57)$$

respectively. Meanwhile, let $A_{\mathbb{C}}$ denote a realization of $-\sum_{i,j=1}^n \frac{\partial}{\partial x_j}\left(a_{ij}(x)\frac{\partial}{\partial x_i}\right)$ in the complex L_2-space $L_2(\Omega; \mathbb{C})$. Then, $A_{\mathbb{C}}$ is a positive definite self-adjoint operator of $L_2(\Omega; \mathbb{C})$ and its domain is, as for A, characterized by (3.10) but $H^2(\Omega; \mathbb{C})$ and $\mathring{H}^1(\Omega; \mathbb{C})$ should be substituted with the corresponding ones. The characterization (3.11) is similarly rewritten as $\mathcal{D}(A_{\mathbb{C}}^{\frac{1}{2}}) = \mathring{H}^1(\Omega; \mathbb{C})$. Moreover, as seen in Subsection 1.4.2, $A_{\mathbb{C}}$ enjoys the relation

$$A_{\mathbb{C}}u = A(\operatorname{Re} u) + iA(\operatorname{Im} u), \qquad u \in \mathcal{D}(A_{\mathbb{C}}),$$

which means that $A_{\mathbb{C}}$ is a real operator of $L_2(\Omega; \mathbb{C})$ and that $[A_{\mathbb{C}}]_{|L_2(\Omega)} = A$. Thereby, identifying A with $A_{\mathbb{C}}$, we can consider A to be a complex linear operator acting in $L_2(\Omega; \mathbb{C})$.

Next, extend $f(x, u)$ to a complex analytic function for suitable complex variables u as follows. Taking account of the assumption (3.13), it is possible to choose a complex convex domain $I_{\mathbb{C}}$ such that

$$\overline{\{u(x,t); \; (x,t) \in \Omega \times [0, \infty)\}} \subset I_{\mathbb{C}}$$

and that

for each $x \in \overline{\Omega}$, $f(x, u)$ is an analytic function for $u \in I_{\mathbb{C}}$.

Furthermore, in view of (3.57), there exists a radius $r > 0$ such that, if $u \in B^{Y_{\mathbb{C}}}(\overline{u}; r)$, then $u(x) \in I_{\mathbb{C}}$ for all $x \in \overline{\Omega}$.

As a consequence, the mapping $u \mapsto \dot{\Phi}(u)$ mentioned in Proposition 3.6 can be extended to a complex form as

$$[\dot{\Phi}]_{\mathbb{C}}u = u - A^{-1}f(x, u), \qquad u \in B^{Y_{\mathbb{C}}}(\overline{u}; r).$$

Proposition 3.7 *The correspondence $u \mapsto [\dot{\Phi}]_{\mathbb{C}}(u)$ maps $B^{Y_{\mathbb{C}}}(\overline{u}; r)$ into $Y_{\mathbb{C}}$ and is continuously Fréchet differentiable. For any $u \in B^{Y_{\mathbb{C}}}(\overline{u}; r)$, the derivative is given by*

$$\left[[\dot{\Phi}]_{\mathbb{C}}\right]'(u)h = h - A^{-1}f_u(x, u)h, \qquad h \in Y_{\mathbb{C}}. \qquad (3.58)$$

Proof If we notice Lemma 3.3 below, then the proof of the proposition is carried out in a way similar to the proof of Proposition 3.6. So the detail will be omitted. □

Lemma 3.3 *It holds that*

$$|f_u(x, v) - f_u(x, u)| + |f(x, u) - f(x, v)| \le C_f |v - u|, \qquad x \in \overline{\Omega}; \ u, \ v \in I_{\mathbb{C}},$$

with some constant $C_f > 0$.

Proof Let $u, \ v \in I_{\mathbb{C}}$. As $I_{\mathbb{C}}$ is convex, we have

$$f_u(x, v) - f_u(x, u) = \int_0^1 f_{uu}(x, \theta v + (1 - \theta)u)(v - u)d\theta.$$

Hence, the first estimate is obtained. It is the same for the second one. $\square$

In view of (3.58), we are led to introduce for the operator L its complex form by

$$L_{\mathbb{C}}h = h - A^{-1} f_u(x, \overline{u})h, \qquad h \in Y_{\mathbb{C}}.$$

Since $f_u(x, \overline{u})$ is real-valued and A^{-1} is real, it naturally follows that

$$L_{\mathbb{C}}h = L(\mathrm{Re}\, h) + iL(\mathrm{Im}\, h), \qquad h \in Y_{\mathbb{C}},$$

which means that $L_{\mathbb{C}}$ is also a real linear operator on $Y_{\mathbb{C}}$ whose real part is the L which was given by (3.49).

Let $h \in \mathcal{K}(L_{\mathbb{C}})$, i.e., $L_{\mathbb{C}}h = 0$; it then follows that $L(\mathrm{Re}\, h) = L(\mathrm{Im}\, h) = 0$; therefore, both $\mathrm{Re}\, h$ and $\mathrm{Im}\, h$ belong to $\mathcal{K}(L)$; namely, we have $\mathcal{K}(L_{\mathbb{C}}) = \mathcal{K}(L) + i\mathcal{K}(L)$. Furthermore, this means that any basis of $\mathcal{K}(L)$ (which consists of real-valued functions) becomes a basis of $\mathcal{K}(L_{\mathbb{C}})$ in $Y_{\mathbb{C}}$, too. In particular, we see that $\dim \mathcal{K}(L_{\mathbb{C}}) = \dim \mathcal{K}(L)$. For the same reason, $h \in L_{\mathbb{C}}(Y_{\mathbb{C}})$ if and only if both $\mathrm{Re}\, h$ and $\mathrm{Im}\, h$ belong to $L(Y)$, i.e., $L_{\mathbb{C}}(Y_{\mathbb{C}}) = L(Y) + iL(Y)$. In particular, $L_{\mathbb{C}}(Y_{\mathbb{C}})$ is a closed subspace of $Y_{\mathbb{C}}$. Then, (3.54) yields that $Y_{\mathbb{C}}$ is topologically decomposed into

$$Y_{\mathbb{C}} = Y + iY = [\mathcal{K}(L) + L(Y)] + i[\mathcal{K}(L) + L(Y)]$$
$$= [\mathcal{K}(L) + i\mathcal{K}(L)] + [L(Y) + iL(Y)] = \mathcal{K}(L_{\mathbb{C}}) + L_{\mathbb{C}}(Y_{\mathbb{C}}). \qquad (3.59)$$

In view of (3.55), set

$$P_{\mathbb{C}}h = P(\mathrm{Re}\, h) + iP(\mathrm{Im}\, h), \qquad h \in Y_{\mathbb{C}}.$$

Then, $P_{\mathbb{C}}$ becomes a complex bounded linear operator on $Y_{\mathbb{C}}$ and is a projection of the decomposition above, namely, $\mathcal{K}(L_{\mathbb{C}}) = P_{\mathbb{C}}(Y_{\mathbb{C}})$ and $L_{\mathbb{C}}(Y_{\mathbb{C}}) = (I - P_{\mathbb{C}})(Y_{\mathbb{C}})$. Of course, $L_{\mathbb{C}}$ is an isomorphism from $L(Y_{\mathbb{C}})$ onto itself.

Identifying L (resp. P) with $L_{\mathbb{C}}$ (resp. $P_{\mathbb{C}}$), let us consider L (resp. P) to be a complex bounded linear operator on $Y_{\mathbb{C}}$.

We are now ready to define a complex critical manifold for S. In fact, we set

$$S_{\mathbb{C}} = \{u \in B^{Y_{\mathbb{C}}}(\overline{u}; r);\ (I - P)[\dot{\Phi}]_{\mathbb{C}}(u) = 0\}$$

in the neighborhood $B^{Y_{\mathbb{C}}}(\overline{u}; r)$. Then, since $Y_{\mathbb{C}}$ is decomposed into the form (3.59), $[\dot{\Phi}]_{\mathbb{C}}(u)$ is continuously Fréchet differentiable with $\left[[\dot{\Phi}]_{\mathbb{C}}\right]'(\overline{u}) = L$ and since $(I - P)L = L$ is an isomorphism of $L(Y_{\mathbb{C}})$, it is possible to repeat the same arguments as in the proof of Preposition 2.2 to claim that there is an open neighborhood $U = U_0 \times U_1$ of $\overline{u}$ in $Y_{\mathbb{C}}$, where U_0 (resp. U_1) is an open neighborhood of $P\overline{u}$ (resp. $(I - P)\overline{u}$) in $\mathcal{K}(L)$ (resp. $L(Y_{\mathbb{C}})$), such that $S_{\mathbb{C}}$ is given in U by

$$S_{\mathbb{C}} \cap U = \{(u_0, g_{\mathbb{C}}(u_0));\ u_0 \in U_0,\ g_{\mathbb{C}} : U_0 \to U_1\},$$

where $g_{\mathbb{C}}$ is a mapping from U_0 into U_1 satisfying $g_{\mathbb{C}}(P\overline{u}) = (I - P)\overline{u}$ and is continuously Fréchet differentiable. Hence, $S_{\mathbb{C}}$ is a complex $\mathcal{C}^1$-manifold having the same dimension as $\mathcal{K}(L)$.

As noticed above, the real basis $v_1, v_2, \ldots, v_N$ of the real $\mathcal{K}(L)$ is still a basis for the complex $\mathcal{K}(L)$ in $Y_{\mathbb{C}}$. So, identify $\mathcal{K}(L)$ with $\mathbb{C}^N$ by the correspondence

$$u_0 = \sum_{k=1}^{N} \zeta_k v_k \in \mathcal{K}(L) \quad \longleftrightarrow \quad \zeta = (\zeta_1, \zeta_2, \ldots, \zeta_N) \in \mathbb{C}^N.$$

Let U_0 correspond to an open neighborhood $\Omega_{\mathbb{C}}$ of $\overline{\xi}$ in $\mathbb{C}^N$. In $\Omega_{\mathbb{C}}$, we want to consider the function

$$\phi_{\mathbb{C}}(\zeta) = \Phi_{\mathbb{C}}\left(\sum_{k=1}^{N} \zeta_k v_k + g_{\mathbb{C}}(\sum_{k=1}^{N} \zeta_k v_k)\right), \qquad \zeta \in \Omega_{\mathbb{C}},$$

introducing a complexification $\Phi_{\mathbb{C}}(u)$ of $\Phi(u)$ given by

$$\Phi_{\mathbb{C}}(u) = \int_{\Omega} \left(\frac{1}{2} \sum_{i,j=1}^{n} a_{ij}(x) \frac{\partial u}{\partial x_i} \frac{\partial u}{\partial x_j} - F(x, u)\right) dx, \qquad u \in B^{Y_{\mathbb{C}}}(\overline{u}; r),$$

where $F(x, u) = \int_{u_*}^{u} f(x, v) dv$ for $u \in I_{\mathbb{C}}$, u_* being the central point of the interval I. Clearly, $F(x, u)$ is also an analytic function for $u \in I_{\mathbb{C}}$. It is then seen that $\phi_{\mathbb{C}}(\zeta)$ is continuously differentiable for each complex variable ζ_k. The characterization of analytic functions of several complex variables (see [Die60, (9.10.1)] or [Hör90, Theorem 2.2.8]) is now available to $\phi_{\mathbb{C}}(\zeta)$ to conclude its analyticity in $\Omega_{\mathbb{C}}$. As a consequence, (3.56) has been proved.

In this way, we know that the assumption (2.35) is fulfilled. Theorem 2.2 ultimately provides that there exists an exponent $0 < \theta \leq \frac{1}{2}$ for which it holds that

$$\|\dot{\Phi}(u)\|_Y \geq C |\Phi(u) - \Phi(\overline{u})|^{1-\theta}, \qquad u \in B^Y(\overline{u}; r), \tag{3.60}$$

with some radius $r > 0$ and constant $C > 0$.

3.6.4 Verification of (2.14)

Finally, on the basis of (3.60), let us verify validity of the condition (2.14). It is in fact possible to use the techniques suggested in Subsection 2.3.5.

(I) *Case where* $n = 2$. From (3.9), (3.47) and (3.50), we see that $Z \subset Y$. Take an exponent p' so that $2 < p' < p$. Then, since $H_{p'}^1(\Omega) \subset \mathcal{C}(\overline{\Omega})$ and since

$\|u\|_{L_{p'}} \leq \|u\|_{L_p}^{\alpha} \|u\|_{L_2}^{1-\alpha}$ for $u \in L_p(\Omega)$ with $\alpha = \left(\frac{1}{2} - \frac{1}{p'}\right) / \left(\frac{1}{2} - \frac{1}{p}\right)$ by the

Hölder inequality, it follows that $\|u\|_Y \leq C\|u\|_{H_{p'}^1} \leq C\|u\|_{H_p^1}^{\alpha} \|u\|_{H^1}^{1-\alpha}$, $u \in$

$H_p^1(\Omega) \cap X$. Therefore, there exists a radius $r''' > 0$ such that $\overline{B}^{H_p^1}(0; C_{1,p}) \cap$ $B^X(\overline{u}; r''') \subset B^Y(\overline{u}; r)$, $C_{1,p}$ being the constant appearing in (3.17) and r being the radius appearing in (3.60). As $\|\dot{\Phi}(u)\|_Y \leq C\|\dot{\Phi}(u)\|_Z$, we conclude that (2.14) is fulfilled.

(II) *Case where* $n = 3$. From (3.10), (3.47) and (3.50), in this case, too, we see that $Z \subset Y$. Take an exponent s so that $\frac{3}{2} < s < 2$. Then, since $H^s(\Omega) \subset$ $\mathcal{C}(\overline{\Omega})$ and since $\|u\|_{H^s} \leq C\|u\|_{H^2}^{\alpha} \|u\|_{H^1}^{1-\alpha}$ for $u \in H^2(\Omega)$ with $\alpha = s - 1$, it follows that $\|u\|_Y \leq C\|u\|_{H^2}^{\alpha} \|u\|_{H^1}^{1-\alpha}$, $u \in Z$. Therefore, there exists a radius $r''' > 0$ such that $\overline{B}^{H^2}(0; C_{2,2}) \cap B^X(\overline{u}; r''') \subset B^Y(\overline{u}; r)$, $C_{2,2}$ being the constant appearing in (3.19) and r being the radius appearing in (3.60). As $\|\dot{\Phi}(u)\|_Y \leq C\|\dot{\Phi}(u)\|_Z$, (2.14) is fulfilled.

(III) *Case where* $n = 4$. We have to notice that it holds true that $H^s(\Omega) \subset$ $\mathcal{C}(\overline{\Omega})$ only for $s > 2$ but not for $s = 2$, which means that $Z \subset Y$ fails now. So, let $2 < s < 3$. Then, since $H^s(\Omega) \subset \mathcal{C}(\overline{\Omega})$ and since $\|u\|_{H^s} \leq C\|u\|_{H^3}^{\alpha} \|u\|_{H^1}^{1-\alpha}$ for $u \in H^3(\Omega)$ with $\alpha = \frac{s-1}{2}$, it follows that $\|u\|_Y \leq C\|u\|_{H^3}^{\alpha} \|u\|_{H^1}^{1-\alpha}$, $u \in H^3(\Omega) \cap X$. Therefore, there exists $r''' > 0$ such that $\overline{B}^{H^3}(0; C_{3,2}) \cap B^X(\overline{u}; r''') \subset B^Y(\overline{u}; r)$, $C_{3,2}$ being the constant appearing in (3.20) and r being the radius appearing in (3.60).

Meanwhile, since $\|u\|_{H^s} \leq C\|u\|_{H^3}^{\beta} \|u\|_{H^2}^{1-\beta}$ for $u \in H^3(\Omega)$ with $\beta = s-2$, it follows that $\|\dot{\Phi}(u)\|_Y \leq C\|\dot{\Phi}(u)\|_{H^3}^{\beta} \|\dot{\Phi}(u)\|_{H^2}^{1-\beta}$ for $u \in H^3(\Omega) \cap Z$. So, $\|\dot{\Phi}(u)\|_Y \leq C\|\dot{\Phi}(u)\|_Z^{1-\beta}$ for $u \in \overline{B}^{H^3}(0; C_{3,2}) \cap Z$. This together with (3.60) then yields that

$$C\|\dot{\Phi}(u)\|_Z \geq |\Phi(u) - \Phi(\overline{u})|^{1-\theta'}, \qquad u \in \overline{B}^{H^3}(0; C_{3,2}) \cap B^X(\overline{u}; r'''),$$

with $\theta' = 1 - \frac{1-\theta}{1-\beta}$. If s is taken so that $2 < s < 2+\theta$, then θ' can be a positive exponent. Hence, in view of (3.20), (2.14) is fulfilled.

(IV) *Case where* $n \geq 5$. In view of $H^s(\Omega) \subset \mathcal{C}(\overline{\Omega})$ only for $s > \frac{n}{2}$, fix an exponent s so that $s > \frac{n}{2}$. On the other hand, let m denote an integer such that $s < 2(m + 1)$. Then, since $H^s(\Omega) \subset \mathcal{C}(\overline{\Omega})$ and since $\|u\|_{H^s} \leq C\|u\|_{H^{2(m+1)}}^{\alpha} \|u\|_{H^1}^{1-\alpha}$ for $u \in H^{2(m+1)}(\Omega)$ with $\alpha = \frac{s-1}{2m+1}$, it follows that

$\|u\|_Y \leq C\|u\|_{H^{2(m+1)}}^{\alpha}\|u\|_{H^1}^{1-\alpha}$, $u \in H^{2(m+1)}(\Omega) \cap X$. Therefore, there exists $r''' > 0$ such that $\overline{B}^{H^{2(m+1)}}(0; D_m) \cap B^X(\overline{u}; r''') \subset B^Y(\overline{u}; r)$, D_m being the constant appearing in (3.41) and r being the radius appearing in (3.60).

Meanwhile, since $\|u\|_{H^s} \leq C\|u\|_{H^{2(m+1)}}^{\beta}\|u\|_{H^2}^{1-\beta}$ for $u \in H^{2(m+1)}(\Omega)$ with $\beta = \frac{s-2}{2m}$, it follows that $\|\dot{\Phi}(u)\|_Y \leq C\|\dot{\Phi}(u)\|_{H^{2(m+1)}}^{\beta}\|\dot{\Phi}(u)\|_{H^2}^{1-\beta}$ for $u \in H^{2(m+1)}(\Omega) \cap Z$. Therefore, $\|\dot{\Phi}(u)\|_Y \leq C\|\dot{\Phi}(u)\|_Z^{1-\beta}$ for $u \in \overline{B}^{H^{2(m+1)}}(0; D_m) \cap Z$. This together with (3.60) then yields that

$$C\|\dot{\Phi}(u)\|_Z \geq |\Phi(u) - \Phi(\overline{u})|^{1-\theta'}, \qquad u \in \overline{B}^{H^{2(m+1)}}(0; D_m) \cap B^X(\overline{u}; r'''),$$

with $\theta' = 1 - \frac{1-\theta}{1-\beta}$. Hence, if m is taken sufficiently large so that $m > \frac{s-2}{2\theta}$, then θ' can be a positive exponent; consequently, in view of (3.41), (2.14) is fulfilled.

We are now ready to apply Theorem 2.1 to (3.7) or (3.1). In fact, we obtain the following results. Assume that: when $n = 2$, Ω is a bounded domain with Lipschitz boundary, $a_{ij}(x)$ are in $L_\infty(\Omega)$ and $f(x, u)$ belongs to $\mathcal{C}^{0,\infty}(\overline{\Omega} \times \mathbb{R})$; when $n = 3$, Ω is a convex or $\mathcal{C}^2$ bounded domain, $a_{ij}(x)$ are in $\mathcal{C}^1(\overline{\Omega})$ and $f(x, u)$ belongs to $\mathcal{C}^{0,\infty}(\overline{\Omega} \times \mathbb{R})$; when $n = 4$, Ω is a $\mathcal{C}^3$ bounded domain, $a_{ij}(x)$ are in $\mathcal{C}^2(\overline{\Omega})$ and $f(x, u)$ belongs to $\mathcal{C}^{1,\infty}(\overline{\Omega} \times \mathbb{R})$; and when $n \geq 5$, Ω is a $\mathcal{C}^\infty$ bounded domain, $a_{ij}(x)$ are in $\mathcal{C}^\infty(\overline{\Omega})$ and $f(x, u)$ belongs to $\mathcal{C}^{\infty,\infty}(\overline{\Omega} \times \mathbb{R})$. In addition, assume (3.3) and (3.4). Let there exist a global solution $u(t)$ to (3.7) lying in (3.12) and let the analyticity condition (3.13) be satisfied. Then, as $t \to \infty$, the solution $u(t)$ converges to a stationary solution $\overline{u}$ of (3.7) at a rate

$$\|u(t) - \overline{u}\|_{L_2} \leq (D\theta'')^{-1}[\Phi(u(t)) - \Phi(\overline{u})]^{\theta''} \qquad \text{for all sufficiently large } t.$$

Here, when $n = 2, 3$, θ'' denotes the exponent θ obtained above; meanwhile, when $n \geq 4$, it denotes the exponent θ'. Also D is a constant appearing in (2.14).

3.7 Reaction–Diffusion Equations

Consider the reaction–diffusion equations

$$\begin{cases} \dfrac{\partial u}{\partial t} - a\Delta u = f(u, v) & \text{in } \Omega \times (0, \infty), \\[2mm] \dfrac{\partial v}{\partial t} - b\Delta v = g(u, v) & \text{in } \Omega \times (0, \infty), \\[2mm] u = v = 0 & \text{on } \partial\Omega \times (0, \infty), \end{cases} \qquad (3.61)$$

in a convex or $\mathcal{C}^2$, bounded domain $\Omega \subset \mathbb{R}^3$. Here, $u = u(x, t)$ and $v = v(x, t)$ are unknown functions satisfying the homogeneous Dirichlet conditions on $\partial\Omega$, $a > 0$

and $b > 0$ are diffusion constants, and $f(u, v)$ and $g(u, v)$ denote reaction functions defined for $(u, v) \in \mathbb{R}^2$.

We assume that (3.61) is of divergence form, namely, there is a real smooth function $F(u, v)$ such that

$$f(u, v) = F_u(u, v) \quad \text{and} \quad g(u, v) = F_v(u, v) \qquad \text{for } (u, v) \in \mathbb{R}^2. \qquad (3.62)$$

Equations (3.61) can be handled as an abstract evolution equation

$$U' + \mathbb{A}U = F'(U), \qquad 0 < t < \infty, \qquad (3.63)$$

in a product L_2-space

$$\mathbb{L}_2(\Omega) = \left\{ U = \begin{pmatrix} u \\ v \end{pmatrix}; \ u \in L_2(\Omega) \quad \text{and} \quad v \in L_2(\Omega) \right\}.$$

Here, $\mathbb{A} = \begin{pmatrix} A & 0 \\ 0 & B \end{pmatrix}$ is an operator matrix such that A (resp. B) is a realization of $-a\Delta$ (resp. $-b\Delta$) in $L_2(\Omega)$ under the homogeneous Dirichlet boundary conditions. Then, A and B are positive definite self-adjoint operators of $L_2(\Omega)$ with the domains $\mathcal{D}(A) = \mathcal{D}(B) = H^2(\Omega) \cap \overset{\circ}{H}{}^1(\Omega)$ (see (3.10)), and are real sectorial operators (see Subsection 1.4.2). Consequently, $\mathbb{A}$ is a positive definite self-adjoint operator of $\mathbb{L}_2(\Omega)$ with $\mathcal{D}(\mathbb{A}) = \mathbb{H}^2(\Omega) \cap \overset{\circ}{\mathbb{H}}{}^1(\Omega)$, and is a real sectorial operator. In addition, (3.11) implies that $\mathcal{D}(\mathbb{A}^{\frac{1}{2}}) = \overset{\circ}{\mathbb{H}}{}^1(\Omega)$. Meanwhile, $F'(U)$ is a nonlinear operator of $\mathbb{L}_2(\Omega)$ given by

$$F'(U) = \begin{pmatrix} F_u(u, v) \\ F_v(u, v) \end{pmatrix} = \begin{pmatrix} f(u, v) \\ g(u, v) \end{pmatrix}, \qquad U = \begin{pmatrix} u \\ v \end{pmatrix}; \ u, v \in \mathcal{C}(\overline{\Omega}).$$

As the prerequisite assumption, let there exist a global solution $U(t) = \begin{pmatrix} u(t) \\ v(t) \end{pmatrix}$ to (3.63) in the function space

$$U \in \mathcal{C}([0, \infty); \mathcal{D}(\mathbb{A})) \cap \mathcal{C}^1([0, \infty); \mathbb{L}_2(\Omega)) \cap \mathcal{B}([0, \infty); \mathbb{L}_\infty(\Omega)). \qquad (3.64)$$

We assume that there exists a two-dimensional domain $\mathbb{I}$ such that

$$\overline{\{(u(x, t), v(x, t)); \ (x, t) \in \Omega \times [0, \infty)\}} \subset \mathbb{I}$$

and that

$$F(u, v) \text{ is analytic for } (u, v) \in \mathbb{I}. \qquad (3.65)$$

As it is concerned with asymptotic behavior of $U(t)$, we are allowed to cut off the values $F(u, v)$ for all sufficiently large (u, v), i.e.,

$$F(u, v) \equiv 0 \text{ for all } (u, v) \in \mathbb{R}^2 \text{ if } |u| + |v| \text{ are sufficiently large.} \qquad (3.66)$$

Such a cutoff is not essential, but makes arguments considerably more simple. For example, it follows from (3.66) that $F'(U)$ satisfies a global Lipschitz condition

$$\|F'(U) - F'(V)\|_{\mathbb{L}_2} \leq L\|U - V\|_{\mathbb{L}_2}, \qquad U, V \in \mathbb{L}_2(\Omega), \qquad (3.67)$$

with some constant $L > 0$.

3.7.1 Some Properties of $U(t)$

Let us observe several basic properties of the solution $U(t)$.

First, let us verify the global norm estimate

$$\|U(t)\|_{\mathbb{H}^2} \leq C_2, \qquad \tau_0 \leq \forall t < \infty, \qquad (3.68)$$

with some constant C_2, $\tau_0 > 0$ being any fixed time. But its verification is quite similar to that for (3.19). Indeed, on account of (3.67), it is possible to apply Theorem 1.4 to (3.63) with $\beta = \eta = 0$ to obtain that

$$\|\mathbb{A}U(t)\|_{\mathbb{L}_2} + \|U'(t)\|_{\mathbb{L}_2} \leq C_{U(0)}/t, \qquad 0 < t \leq T_{U(0)},$$

with a time $T_{U(0)} > 0$ and a constant $C_{U(0)} > 0$ which are both determined by magnitude of the norm $\|U(0)\|_{\mathbb{L}_2}$ of initial value $U(0)$ alone. Then, resetting the initial time to any other $\tau > 0$, we will regard $U(t)$ as a solution to (3.63) for $t \in [\tau, \infty)$ with initial value $U(\tau)$. It must then follow that

$$\|\mathbb{A}u(t)\|_{\mathbb{L}_2} + \|U'(t)\|_{\mathbb{L}_2} \leq C_{U(\tau)}/(t - \tau), \qquad \tau < t \leq \tau + T_{U(\tau)},$$

$T_{U(\tau)} > 0$ and $C_{U(\tau)} > 0$ being determined by the norm $\|U(\tau)\|_{\mathbb{L}_2}$ alone. Because of (3.64), $C_{U(\tau)}$ and $T_{U(\tau)}$ are determined uniformly in the initial time τ (> 0); hence, the estimate (3.68) is satisfied.

Second, we see that there is a Lyapunov function for the global solution. Indeed, it is easy to verify that

$$\frac{d}{dt} \int_\Omega F(u(t), v(t))dx = \int_\Omega [F_u(u(t), v(t))u'(t) + F_v(u(t), v(t))v'(t)]dx$$

$$= \int_\Omega [f(u(t), v(t))u'(t) + g(u(t), v(t))v'(t)]dx, \qquad 0 \leq t < \infty.$$

Consider the inner product of the equation of (3.63) and $U'(t)$. Then, we have

$$\|U'(t)\|_{\mathbb{L}_2}^2 + \frac{1}{2}\frac{d}{dt}\int_\Omega [a|\nabla u(t)|^2 + b|\nabla v(t)|^2]dx = \frac{d}{dt}\int_\Omega F(u(t), v(t))dx.$$

Therefore, if $\Phi(U)$ is set as

$$\Phi(U) = \int_\Omega \left(\frac{1}{2}[a|\nabla u|^2 + b|\nabla v|^2] - F(u, v)\right)dx, \qquad U = \begin{pmatrix} u \\ v \end{pmatrix} \in \overset{\circ}{\mathbb{H}}{}^1(\Omega),$$

then it holds true that

$$\frac{d}{dt}\Phi(U(t)) = -\|U'(t)\|_{\mathbb{L}_2}^2, \qquad 0 \le t < \infty, \tag{3.69}$$

which means that $\Phi(U)$ is a Lyapunov function for $U(t)$. It is clear that

$$\lim_{t\to\infty} \Phi(U(t)) = \inf_{0\le t<\infty} \Phi(U(t)) > -\infty. \tag{3.70}$$

Furthermore, due to (3.69),

if $\frac{d}{dt}\Phi(U(t)) = 0$ at $t = \bar{t}$, then $U(\bar{t})$ is a stationary solution of (3.63). (3.71)

Third, let us notice that $U(t)$ has a suitable ω-limit. Let

$$\omega(U) = \{\overline{U} \in \mathbb{L}_2(\Omega); \exists t_m \nearrow \infty \text{ such that } U(t_m) \to \overline{U} \text{ in } \mathbb{L}_2(\Omega)\}.$$

Then, by the same arguments as before, $\omega(U)$ is seen to be a nonempty set, and $\overline{U} \in \omega(U)$ implies that $\overline{U} \in \mathcal{D}(A)$ and that there exists a temporal sequence $t_m \nearrow \infty$ such that $AU(t_m) \to A\overline{U}$ weakly and $U(t_m) \to \overline{U}$ strongly in $\mathbb{L}_2(\Omega)$ respectively. In the meantime, it follows from (3.69) and (3.70) that

$$\int_0^\infty \|U'(t)\|_{\mathbb{L}_2}^2 dt = \Phi(U(0)) - \lim_{t\to\infty} \Phi(U(t)) < \infty.$$

For the same reasons as for (3.46), there is a temporal sequence $t_m \nearrow \infty$ for which it holds that $\|U'(t_m)\|_{\mathbb{L}_2} \to 0$, i.e., $U'(t_m) \to 0$ in $\mathbb{L}_2(\Omega)$. Then, as for Proposition 3.2, we can show that there exists an ω-limit $\overline{U} \in \omega(U)$ which is a stationary solution of (3.63).

3.7.2   Formulation

We know that $U \in \mathcal{C}([0, \infty); \mathcal{D}(A)) \cap \mathcal{C}^1([0, \infty); \mathbb{L}_2(\Omega))$ and that the Lyapunov function for $U(t)$ is defined on $\overset{\circ}{\mathbb{H}}{}^1(\Omega) = \mathcal{D}(A^{\frac{1}{2}})$. By these situations, it is natural to set the triplet $Z \subset X \subset Z^*$ as

$$Z = \mathcal{D}(\mathbb{A}), \quad X = \mathcal{D}(\mathbb{A}^{\frac{1}{2}}), \quad Z^* = \mathcal{D}(\mathbb{A}^0) = \mathbb{L}_2(\Omega). \tag{3.72}$$

Obviously, Z (resp. X) is a Hilbert space equipped with the inner product $(\cdot, \cdot)_Z = (\mathbb{A}\cdot, \mathbb{A}\cdot)_{\mathbb{L}_2}$ (resp. $(\cdot, \cdot)_X = (\mathbb{A}^{\frac{1}{2}}\cdot, \mathbb{A}^{\frac{1}{2}}\cdot)_{\mathbb{L}_2}$). So, the duality product between Z and Z^* is given by

$$\langle U, F \rangle_{Z \times Z^*} = (\mathbb{A}U, F)_{\mathbb{L}_2}, \qquad (U, F) \in Z \times Z^*$$

$$\langle U, V \rangle_{Z \times Z^*} = (\mathbb{A}^{\frac{1}{2}}U, \mathbb{A}^{\frac{1}{2}}V)_{\mathbb{L}_2}, \qquad (U, V) \in Z \times X.$$

Because of (3.68), $U(t)$ then satisfies not only (2.3), but also (2.4).

The Fréchet differentiability of $\Phi : X \to \mathbb{R}$ can be verified as for Proposition 3.3. Indeed, $\Phi(U)$ is continuously Fréchet differentiable in X with the derivative

$$\dot{\Phi}(U) = U - \mathbb{A}^{-1}F'(U), \qquad U \in X. \tag{3.73}$$

Moreover, if $U \in Z$, then $\dot{\Phi}(U) \in Z$ and the mapping $U \mapsto \dot{\Phi}(U)$ is continuous from Z into itself.

As for the ω-limit of $U(t)$, we choose the $\overline{U} \in \omega(U)$ which is a stationary solution of (3.63) whose existence is verified above.

3.7.3 Verification of Structural Assumptions

Let us verify all the structural assumptions announced in Section 2.2.

(I) *Critical Condition.* The ω-limit $\overline{U}$ has been chosen so that $\overline{U}$ is a stationary solution to (3.63). Since $\mathbb{A}\overline{U} = F'(\overline{U})$, (3.73) implies $\dot{\Phi}(\overline{U}) = 0$.

(II) *Lyapunov Function.* By (3.69), $\frac{d}{dt}\Phi(U(t)) \leq 0$ for any $0 \leq t < \infty$. According to (3.71), if $\frac{d}{dt}\Phi(U(t)) = 0$ at some time $t = \overline{t}$, then $U(\overline{t})$ is a stationary solution of (3.63) and $U(t) = U(\overline{t})$ for all $t \geq \overline{t}$. Thereby, Theorem 3.1 below is automatically valid. For this reason, it suffices to argue under (2.12).

(III) *Angle Condition.* On account of (3.72) and (3.73), we can write

$$-\langle \dot{\Phi}(U(t)), U'(t) \rangle_{Z \times Z^*} = -(\mathbb{A}\dot{\Phi}(U(t)), U'(t))_{\mathbb{L}_2}$$

$$= \| - \mathbb{A}U(t) + F'(U(t)) \|_{\mathbb{L}_2}^2 = \| U'(t) \|_{\mathbb{L}_2}^2$$

$$= \| \dot{\Phi}(U(t)) \|_Z \| U'(t) \|_{Z^*},$$

which shows that (2.13) is fulfilled with $\delta = 1$.

(IV) *Gradient Inequality.* Of course, verification of the condition (2.14) is not so immediate. But it is possible to follow the same procedure as for the case of single equation (3.7) where $n = 3$. As for the space Y (see (3.50)), we will choose

$$Y = \overset{\circ}{\mathbb{H}}{}^1(\Omega) \cap \left\{ U = \begin{pmatrix} u \\ v \end{pmatrix}; \ u \text{ and } v \text{ are in } \mathcal{C}(\overline{\Omega}) \right\}.$$

Then, it can be shown by arguments analogous to those for (3.60) that

$$\|\dot{\Phi}(U)\|_Y \geq C|\Phi(U) - \Phi(\overline{U})|^{1-\theta}, \qquad U \in B^Y(\overline{U}; r), \tag{3.74}$$

with some exponent $0 < \theta \leq \frac{1}{2}$, some radius $r > 0$ and some constant C. Since $Z \subset \mathcal{D}(A^{\tilde{\theta}}) \subset Y$ with any exponent $\frac{3}{4} < \tilde{\theta} < 1$, there is a radius $r''' > 0$ such that $B^Z(0; C_2) \cap B^X(\overline{U}; r''') \subset B^Y(\overline{U}; r)$; here, C_2 is the constant appearing in (3.68). Hence, (3.74) yields (2.14).

Thus, by applying Theorem 2.1, we obtain the following result.

Theorem 3.1 *Let $\Omega \subset \mathbb{R}^3$ be a convex or $\mathcal{C}^2$, bounded domain and let the functions $f(u, v)$ and $g(u, v)$ be given as (3.62). Let there exist a global solution $U(t)$ to (3.63) belonging to (3.64) and assume that $F(U)$ satisfies the analyticity condition (3.65). Then, as $t \to \infty$, $U(t)$ converges to a stationary solution $\overline{U}$ of (3.63) at a rate*

$$\|U(t) - \overline{U}\|_{\mathbb{L}_2} \leq (D\theta)^{-1}[\Phi(U(t)) - \Phi(\overline{U})]^{\theta} \qquad \text{for all sufficiently large } t,$$

where D is a constant appearing in (2.14).

3.8   Notes and Future Studies

For the asymptotic convergence of solutions to (3.1), the one-dimensional case is exceptional, see Zelenyak [Zel68] and Matano [Mat78]. According to [Mat78], any bounded solution of (3.1) is convergent to a stationary solution even if the analyticity condition (3.13) is not satisfied. On the contrary, Poláčik–Rybakowski [PR96] and Poláčik–Simondon [PS02] have proved the counter result that: For any $\mathcal{C}^2$ bounded domain $\Omega \subset \mathbb{R}^n$, where $n \geq 2$, one can find a $\mathcal{C}^\infty$ function $f(x, u)$ for which Eq. (3.1), where $a_{ij}(x) = \delta_{ij}$ (Kronecker's delta), has a bounded solution whose ω-limit set is a continuum of equilibria. Therefore, to show any general result guaranteeing the asymptotic convergence in a higher-dimensional case, an analyticity assumption like (3.13) is indispensable.

There are big gaps between the cases of dimension four and dimension $n \geq 5$, in the basic assumptions (3.2), (3.5) and (3.6) concerning the regularities of domain Ω, coefficients $a_{ij}(x)$ and external force function $f(x, u)$, respectively. In the cases of $n \geq 5$, these $\mathbb{C}^\infty$ regularities are not necessary conditions but only sufficient conditions. Accurate orders of regularities which we need in the arguments are determined by the exponent θ appearing in (3.60). However, it is not possible to know precisely an optimal value of such θ for given Ω, $a_{ij}(x)$ and $f(x, u)$.

Many authors have already studied the asymptotic convergence problem for (3.1) in higher-dimensional cases, including Jendoubi [Jen98], Chill [Chi03], Chill–Haraux–Jendoubi [CHJ09], and so on. They have proved the asymptotic convergence under some technical assumptions other than (3.12) or (3.13). Then, the results explained in this chapter just show that those assumptions are all cleared up under the regularities (3.2), (3.5) and (3.6) by means of the techniques of abstract parabolic evolution equations.

Concerning the reaction–diffusion equations (3.61), we handled only the simplest ones. It is possible to extend the results obtained here to the general reaction–diffusion equations

$$
\begin{cases}
\dfrac{\partial u_1}{\partial t} - \dfrac{\partial}{\partial x_i}\left(a_{1;i,j}(x)\dfrac{\partial u_1}{\partial x_j}\right) = F_{u_1}(x; u_1, u_2, \ldots, u_m) & \text{in } \Omega \times (0, \infty), \\[3mm]
\dfrac{\partial u_2}{\partial t} - \dfrac{\partial}{\partial x_i}\left(a_{2;i,j}(x)\dfrac{\partial u_2}{\partial x_j}\right) = F_{u_2}(x; u_1, u_2, \ldots, u_m) & \text{in } \Omega \times (0, \infty), \\[3mm]
\quad\cdots\cdots\cdots\cdots \\[1mm]
\dfrac{\partial u_m}{\partial t} - \dfrac{\partial}{\partial x_i}\left(a_{m;i,j}(x)\dfrac{\partial u_m}{\partial x_j}\right) = F_{u_m}(x; u_1, u_2, \ldots, u_m) & \text{in } \Omega \times (0, \infty), \\[3mm]
u_1 = u_2 = \cdots = u_m = 0 & \text{on } \partial\Omega \times (0, \infty),
\end{cases}
$$

in a bounded domain $\Omega \subset \mathbb{R}^n$, where $n = 1, 2, \ldots$, under assumptions analogous to (3.2), (3.3)–(3.5), (3.6) and (3.65) for Ω, $a_{k;i,j}(x)$ and $F(x; u_1, u_2, \ldots, u_m)$, respectively.

The results of this chapter for the three-dimensional equation (3.1) were already published in Iwasaki–Yagi [IY]. Other results are new.

Chapter 4
Epitaxial Growth Model

This chapter is devoted to considering an epitaxial growth model in surface science. The model equation includes an effect of surface diffusion which is described using a biharmonic operator. Under suitable assumptions, we shall show that the results reviewed in Chap. 2 are available to the model equation.

4.1 Model Equation of Fourth Order

Consider an epitaxial growth equation

$$
\begin{cases}
\dfrac{\partial u}{\partial t} + a\Delta^2 u = -\nabla \cdot \left[q(|\nabla u|^2)\nabla u \right] & \text{in} \quad \Omega \times (0, \infty), \\[2mm]
u = \dfrac{\partial u}{\partial n} = 0 & \text{on} \quad \partial\Omega \times (0, \infty),
\end{cases}
\tag{4.1}
$$

in a convex-polygonal or $\mathcal{C}^4$, bounded domain $\Omega \subset \mathbb{R}^2$. Here, $a > 0$ is a surface diffusion constant, $q(\xi) \geq 0$ is a given function for $0 \leq \xi < \infty$ of class $\mathcal{C}^3$, and $u = u(x, t)$ is an unknown function of $(x, t) \in \Omega \times [0, \infty)$ satisfying the homogeneous Dirichlet conditions of second order on $\partial\Omega$.

Johnson–Orme–Hunt–Graff–Sudijono–Sauder–Orr [JOHGSSO94] presented Eq. (4.1) for describing the growing process of a crystal surface under molecular beam epitaxy. The unknown function $u(x, t)$ denotes a displacement of surface height from the standard level at position $x \in \Omega$ and time $t \in [0, \infty)$.

© The Author(s), under exclusive license to Springer Nature Singapore Pte Ltd. 2021
A. Yagi, *Abstract Parabolic Evolution Equations and Łojasiewicz–Simon Inequality II*,
SpringerBriefs in Mathematics, https://doi.org/10.1007/978-981-16-2663-0_4

We will formulate (4.1) into the abstract equation

$$u' + Au = f(u), \qquad 0 < t < \infty, \tag{4.2}$$

in the space $L_2(\Omega)$. Here, A is a realization of the biharmonic operator $a\Delta^2$ in $L_2(\Omega)$ under the Dirichlet conditions $u = \frac{\partial u}{\partial n} = 0$ on $\partial\Omega$ with domain $\mathcal{D}(A) \subset H^3(\Omega)$. We shall describe its precise definition and investigate its basic properties in the next section. Meanwhile, $f(u)$ is a nonlinear operator of $L_2(\Omega)$ defined by

$$
\begin{aligned}
f(u) &= -\nabla \cdot [q(|\nabla u|^2)\nabla u] \\
&= -[q'(|\nabla u|^2)\nabla(|\nabla u|^2) \cdot \nabla u + q(|\nabla u|^2)\Delta u], \qquad u \in H^2(\Omega) \cap \mathcal{C}^1(\overline{\Omega}).
\end{aligned}
\tag{4.3}
$$

On account of (4.3), $f(u)$ is seen to be linear with respect to the second-order partial derivatives of u. Therefore, $f(u)$ satisfies a Lipschitz condition of the form

$$
\begin{aligned}
\|f(u) - f(v)\|_{L_2} \le{}& \psi(\|\nabla u\|_{\mathcal{C}} + \|\nabla v\|_{\mathcal{C}})[\|u - v\|_{H^2} \\
&+ (\|u\|_{H^2} + \|v\|_{H^2})\|\nabla[u - v]\|_{\mathcal{C}}], \qquad u, v \in H^2(\Omega) \cap \mathcal{C}^1(\overline{\Omega}),
\end{aligned}
$$

in $L_2(\Omega)$, $\psi(\cdot)$ being a suitable continuous increasing function determined from $q(\xi)$. Fix an exponent β arbitrarily so that $\frac{1}{2} < \beta < 1$; then, by the embedding (4.14) or (4.15) below, it holds that $\mathcal{D}(A^\beta) \subset H^{2\beta+1}(\Omega) \subset H^2(\Omega) \cap \mathcal{C}^1(\overline{\Omega})$. Consequently, $f(u)$ satisfies the Lipschitz condition

$$\|f(u) - f(v)\|_{L_2} \le \widetilde{\psi}(\|A^\beta u\|_{L_2} + \|A^\beta v\|_{L_2})\|A^\beta(u - v)\|_{L_2}, \qquad u, v \in \mathcal{D}(A^\beta), \tag{4.4}$$

$\widetilde{\psi}(\cdot)$ being a suitable continuous increasing function.

It is then possible to apply Theorem 1.4 to obtain local existence of solutions. Indeed, in view of (4.4), for any $u(0) \in \mathcal{D}(A^\beta)$, there exists a unique local solution $u(t)$ to (4.2) in the function space

$$u \in \mathcal{C}([0, T_{u(0)}]; \mathcal{D}(A^\beta)) \cap \mathcal{C}((0, T_{u(0)}]; \mathcal{D}(A)) \cap \mathcal{C}^1((0, T_{u(0)}]; L_2(\Omega)),$$

and $u(t)$ satisfies the estimate

$$\|Au(t)\|_{L_2} \le C_{u(0)}t^{\beta-1}, \qquad 0 < t \le T_{u(0)}, \tag{4.5}$$

where the time $T_{u_0} > 0$ and the constant $C_{u_0} > 0$ are determined by the magnitude of norm $\|A^\beta u(0)\|_{L_2}$ alone.

As the prerequisite assumption, let there exist a global solution $u(t)$ of (4.2) for $u(0) \in \mathcal{D}(A)$ belonging to the function space

$$u \in \mathcal{C}([0, \infty); \mathcal{D}(A)) \cap \mathcal{C}^1([0, \infty); L_2(\Omega)), \quad |\nabla u| \in \mathcal{B}([0, \infty); \mathcal{C}(\overline{\Omega})). \quad (4.6)$$

We want to show that, as $t \to \infty$, $u(t)$ converges to a stationary solution of (4.2).

As we are concerned with the asymptotic behavior of $u(t)$ only, we are allowed to cut off the values of $q(\xi)$ for sufficiently large ξ. In fact, we assume that

$$q(\xi) \equiv 0 \quad \text{for} \quad \xi \geq R + 1, \quad \text{where} \quad R = \sup_{(x,t) \in \Omega \times [0,\infty)} |\nabla u(x, t)|^2. \quad (4.7)$$

We finally make a crucial assumption that

$$q(\xi) \text{ is analytic for } \xi \text{ in an open interval } I \text{ such that } [0, R] \subset I. \quad (4.8)$$

4.2 Realization of Biharmonic Operator

It is known, according to [Gri85, Corollary 1.5.1.6], that in any $\mathcal{C}^2$ bounded domain Ω, $u \in H^2(\Omega)$ satisfies $u = \frac{\partial u}{\partial n} = 0$ on $\partial \Omega$ if and only if $u \in \overset{\circ}{H}{}^2(\Omega)$. In view of this fact, we choose the space $\overset{\circ}{H}{}^2(\Omega)$ as an underlying space for the unknown function u. Notice that even if Ω is convex-polygonal (i.e., non-smooth), $u \in \overset{\circ}{H}{}^2(\Omega)$ implies $u = \frac{\partial u}{\partial n} = 0$ on $\partial \Omega$.

Let $\overset{\circ}{H}{}^2(\Omega) \subset L_2(\Omega) \subset H^{-2}(\Omega)$ be a triplet of spaces. On $\overset{\circ}{H}{}^2(\Omega)$, we consider a symmetric sesquilinear form

$$a(u, v) = a \int_\Omega \Delta u \cdot \Delta v \, dx, \quad u, v \in \overset{\circ}{H}{}^2(\Omega).$$

By the Poincaré inequality, it holds that $\|u\|_{L_2} \leq C\|\Delta u\|_{L_2}$ for all $u \in H^2(\Omega) \cap \overset{\circ}{H}{}^1(\Omega)$, which provides the conditions (1.43) for this form $a(u, v)$. It is obvious that $a(u, v)$ is a real sesquilinear form defined in Subsection 1.4.1. By the same procedure as in Subsection 1.4.2, $a(u, v)$ determines a linear isomorphism $\mathcal{A}$ from $\overset{\circ}{H}{}^2(\Omega)$ onto $H^{-2}(\Omega)$ by the formula $a(u, v) = \langle \mathcal{A}u, v \rangle_{H^{-2} \times \overset{\circ}{H}{}^2}$.

The operator $\mathcal{A}$ is then considered as a realization of the biharmonic operator $a\Delta^2$ in $H^{-2}(\Omega)$ under the second-order homogeneous Dirichlet boundary conditions with domain $\mathcal{D}(\mathcal{A}) = \overset{\circ}{H}{}^2(\Omega)$. Its part $A \, (= \mathcal{A}_{|L_2})$ in $L_2(\Omega)$ is defined by

$$\begin{cases} \mathcal{D}(A) = \{u \in \overset{\circ}{H}{}^2(\Omega); \, \mathcal{A}u \in L_2(\Omega)\}, \\ Au = \mathcal{A}u. \end{cases} \quad (4.9)$$

This A is considered to be a realization of $a\Delta^2$ in $L_2(\Omega)$ under the same boundary conditions as $\mathcal{A}$. It is seen that A is a positive definite self-adjoint operator of $L_2(\Omega)$. According to (1.46), the domain $\mathcal{D}(A^{\frac{1}{2}})$ of square root $A^{\frac{1}{2}}$ coincides with the definition domain of the form $a(\cdot, \cdot)$; therefore, $\mathcal{D}(A^{\frac{1}{2}}) = \overset{\circ}{H}{}^2(\Omega)$ together with

$$a(u, u) = \|A^{\frac{1}{2}}u\|^2_{L_2}, \qquad u \in \overset{\circ}{H}{}^2(\Omega). \tag{4.10}$$

Furthermore,

$$\langle \varphi, v \rangle_{H^{-2} \times \overset{\circ}{H}{}^2} = \left\langle \mathcal{A}\mathcal{A}^{-1}\varphi, v \right\rangle_{H^{-2} \times \overset{\circ}{H}{}^2}$$
$$= a(\mathcal{A}^{-1}\varphi, v) = (A^{\frac{1}{2}}\mathcal{A}^{-1}\varphi, A^{\frac{1}{2}}v)_{L_2}, \qquad \varphi \in H^{-2}(\Omega), \; v \in \overset{\circ}{H}{}^2(\Omega). \tag{4.11}$$

As for the domain $\mathcal{D}(A)$, we have the following proposition.

Proposition 4.1 *When Ω is convex-polygonal, $\mathcal{D}(A)$ is contained in $H^3(\Omega) \cap \overset{\circ}{H}{}^2(\Omega)$ with the continuous embedding*

$$\|u\|_{H^3} \leq C\|Au\|_{H^{-1}} \leq C\|Au\|_{L_2}, \qquad u \in \mathcal{D}(A). \tag{4.12}$$

Meanwhile, when Ω is of class $\mathcal{C}^4$, $\mathcal{D}(A)$ is characterized as $\mathcal{D}(A) = H^4(\Omega) \cap \overset{\circ}{H}{}^2(\Omega)$ with norm equivalence. In particular,

$$\|u\|_{H^4} \leq C\|Au\|_{L_2}, \qquad u \in \mathcal{D}(A). \tag{4.13}$$

Proof For the case where Ω is convex-polygonal, we can utilize Corollary 7.3.2.5 of Grisvard [Gri85] which gives that $\mathcal{A}$ is an isomorphism from $H^3(\Omega) \cap \overset{\circ}{H}{}^2(\Omega)$ onto $H^{-1}(\Omega)$. In particular, it follows that $\mathcal{D}(A) \subset H^3(\Omega) \cap \overset{\circ}{H}{}^2(\Omega)$ together with the inequality (4.12).

For the case where Ω is of class $\mathcal{C}^4$, we can appeal to the theory of higher order elliptic equations. Among others, the results of [Tan75, Section 3.8] (cf. also [Tan97, Section 5.2]) give that for any $f \in L_2(\Omega)$, there exists a unique solution $u \in H^4(\Omega)$ to the biharmonic equation $a\Delta^2 u = f$ under the Dirichlet boundary conditions $u = \frac{\partial u}{\partial n} = 0$ on $\partial\Omega$, together with the inequality $\|u\|_{H^4} \leq C\|f\|_{L_2}$. This means that $\mathcal{D}(A) \subset H^4(\Omega) \cap \overset{\circ}{H}{}^2(\Omega)$ and (4.13) holds true. Conversely, $u \in H^4(\Omega) \cap \overset{\circ}{H}{}^2(\Omega)$ implies that $a(u, v) = (a\Delta^2 u, v)_{L_2}$ for all $v \in \overset{\circ}{H}{}^2(\Omega)$, i.e., $u \in \mathcal{D}(A)$. $\qquad\square$

From this proposition and (4.10), we verify some results concerning the domains $\mathcal{D}(A^\theta)$ of the fractional powers of A.

Proposition 4.2 *When Ω is convex-polygonal, we have*

$$\mathcal{D}(A^\theta) \subset H^{2\theta+1}(\Omega) \cap \overset{\circ}{H}{}^2(\Omega) \qquad \textit{for any } \tfrac{1}{2} < \theta < 1, \tag{4.14}$$

with the continuous embedding

$$\|u\|_{H^{2\theta+1}} \leq C\|A^\theta u\|_{L_2}, \qquad u \in \mathcal{D}(A^\theta).$$

Meanwhile, when Ω is of class $\mathcal{C}^4$,

$$\mathcal{D}(A^\theta) = H^{4\theta}(\Omega) \cap \overset{\circ}{H}{}^2(\Omega) \qquad \text{for any } \tfrac{1}{2} < \theta < 1, \tag{4.15}$$

with norm equivalence. In particular, we have

$$\|u\|_{H^{4\theta}} \leq C\|A^\theta u\|_{L_2}, \qquad u \in \mathcal{D}(A^\theta).$$

Proof By virtue of (1.41) ($\theta_0 = \tfrac{1}{2}$ and $\theta_1 = 1$), we observe that $\mathcal{D}(A^\theta) = [\mathcal{D}(A^{\frac{1}{2}}), \mathcal{D}(A)]_{2\theta-1}$ for any $\tfrac{1}{2} < \theta < 1$.

Therefore, when Ω is convex-polygonal, we obtain that

$$\mathcal{D}(A^\theta) \subset [\overset{\circ}{H}{}^2(\Omega), H^3(\Omega) \cap \overset{\circ}{H}{}^2(\Omega)]_{2\theta-1}$$

$$\subset [H^2(\Omega), H^3(\Omega)]_{2\theta-1} \cap \overset{\circ}{H}{}^2(\Omega) \subset H^{2\theta+1}(\Omega) \cap \overset{\circ}{H}{}^2(\Omega).$$

As all the embeddings are continuous, we obtain the embedding inequality as well.

Similarly, when Ω is of class $\mathcal{C}^4$, we obtain that

$$\mathcal{D}(A^\theta) \subset [\overset{\circ}{H}{}^2(\Omega), H^4(\Omega) \cap \overset{\circ}{H}{}^2(\Omega)]_{2\theta-1}$$

$$\subset [H^2(\Omega), H^4(\Omega)]_{2\theta-1} \cap \overset{\circ}{H}{}^2(\Omega) \subset H^{4\theta}(\Omega) \cap \overset{\circ}{H}{}^2(\Omega),$$

together with the embedding inequality. In order to prove conversely that $H^{4\theta}(\Omega) \cap \overset{\circ}{H}{}^2(\Omega) \subset \mathcal{D}(A^\theta)$, however, some delicate arguments as in the proof of [Yag10, Theorem 16.12] are required. But, as we need here only the one-side continuous inclusion proved above, its proof may be omitted. □

4.3 Global Boundedness of Sobolev Norms

Our goal is to prove under (4.6) that $u(t)$ satisfies the estimate

$$\|Au(t)\|_{L_2} \leq C_1, \qquad \tau_0 \leq \forall t < \infty, \tag{4.16}$$

with some constant C_1, $\tau_0 > 0$ being any fixed time. The proof consists of four steps.

Step 1 Let us first prove that

$$\|u(t)\|_{L_2} \le C_0, \qquad 0 \le t < \infty, \tag{4.17}$$

with some constant C_0. Indeed, consider the inner product of the equation of (4.2) and $u(t)$. Then,

$$\tfrac{1}{2}\tfrac{d}{dt}\|u(t)\|_{L_2}^2 + \|A^{\frac{1}{2}}u(t)\|_{L_2}^2 = (f(u(t)), u(t))_{L_2}.$$

Here, since

$$(f(u(t)), u(t))_{L_2} = \int_{\Omega} q(|\nabla u(t)|^2)|\nabla u(t)|^2 dx,$$

it follows from (4.6) that $|(f(u(t)), u(t))| \le C$. Hence,

$$\tfrac{1}{2}\tfrac{d}{dt}\|u(t)\|_{L_2}^2 + C^{-1}\|u(t)\|_{L_2}^2 \le C.$$

Solving this differential inequality, we obtain (4.17).

Step 2 Let us next prove that

$$\|A^{\frac{1}{2}}u(t)\|_{L_2} \le C_{\frac{1}{2}}, \qquad 0 \le t < \infty, \tag{4.18}$$

with some constant $C_{\frac{1}{2}}$. Consider now the inner product of the equation of (4.2) and $Au(t)$. Then,

$$\tfrac{1}{2}\tfrac{d}{dt}\|A^{\frac{1}{2}}u(t)\|_{L_2}^2 + \|Au(t)\|_{L_2}^2 = (f(u(t)), Au(t))_{L_2}.$$

Since we can estimate

$$\|f(u(t))\|_{L_2} \le C\|u(t)\|_{H^2} \le C\|A^{\frac{1}{2}}u(t)\|_{L_2} \le C\|u(t)\|_{L_2}^{\frac{1}{2}}\|Au(t)\|_{L_2}^{\frac{1}{2}} \le C\|Au(t)\|_{L_2}^{\frac{1}{2}}$$

due to (4.3), (4.6), (4.10) and (4.17), it follows that $|(f(u(t), Au(t))| \le C\|Au(t)\|_{L_2}^{\frac{3}{2}}$. Therefore, $\tfrac{d}{dt}\|A^{\frac{1}{2}}u(t)\|_{L_2}^2 + \|Au(t)\|_{L_2}^2 \le C$. Consequently, we arrive at

$$\tfrac{d}{dt}\|A^{\frac{1}{2}}u(t)\|_{L_2}^2 + C^{-1}\|A^{\frac{1}{2}}u(t)\|_{L_2}^2 \le C.$$

Solving this differential inequality, (4.18) is obtained.

Step 3 Let us thirdly prove that

$$\|A^{\beta}u(t)\|_{L_2} \le C_{\beta}, \qquad 0 \le t < \infty, \tag{4.19}$$

with some constant C_β (remember that β is fixed above so that (4.4) is valid). We utilize the formula $u(t) = e^{-tA}u(0) + \int_0^t e^{-(t-s)A} f(u(s))ds$ to write

$$A^\beta u(t) = e^{-tA} A^\beta u(0) + \int_0^t A^\beta e^{-(t-s)A} f(u(s))ds.$$

Since $\|f(u(t))\|_{L_2} \le C\|u(t)\|_{H^2} \le CC_{\frac{1}{2}}$ due to (4.18), we can estimate it as

$$\|A^\beta u(t)\|_{L_2} \le e^{-\delta t}\|A^\beta u(0)\|_{L_2} + CC_{\frac{1}{2}} \int_0^t [(t-s)^{-\beta} + 1]e^{-\delta(t-s)}ds.$$

Hence, (4.19) is obtained.

Step 4 As noticed from (4.5), we know that (4.16) is the case in some interval $(0, T_{u(0)}]$. Then, resetting the initial time to any other $\tau > 0$, we regard $u(t)$ as a solution to (4.2) for $t \in [\tau, \infty)$ with initial value $u(\tau)$. Then, it must follow that

$$\|Au(t)\|_{L_2} \le C_{u(\tau)}(t-\tau)^{\beta-1}, \qquad \tau < t \le \tau + T_{u(\tau)},$$

$T_{u(\tau)} > 0$ and $C_{u(\tau)} > 0$ being determined by the norm $\|A^\beta u(\tau)\|_{L_2}$ alone. Whereas, (4.19) shows that this norm is uniformly bounded for $0 \le \tau < \infty$. Hence, (4.16) has been proved. $\qquad\square$

4.4 Some Other Properties of $u(t)$

4.4.1 Lyapunov Function

Let us observe that the following function

$$\Phi(u) = \frac{1}{2} \int_\Omega \left[a|\Delta u|^2 - Q(|\nabla u|^2) \right] dx, \qquad u \in \mathring{H}^2(\Omega), \tag{4.20}$$

is a Lyapunov function for the solution $u(t)$, where $Q(\xi) = \int_0^\xi q(\xi)d\xi$.

Proposition 4.3 *The function* $t \mapsto \int_\Omega Q(|\nabla u(t)|^2)dx$ *is continuously differentiable for* $0 \le t < \infty$ *with the derivative* $2(f(u(t)), u'(t))_{L_2}$.

Proof Let $u, h \in \mathcal{D}(A)$. Then, for each $x \in \Omega$, we have

$$Q(|\nabla[u(x) + h(x)]|^2) - Q(|\nabla u(x)|^2) = \int_0^1 \frac{d}{d\theta} Q(|\nabla[u(x) + \theta h(x)]|^2)d\theta$$

$$= 2 \int_0^1 q(|\nabla[u(x) + \theta h(x)]|^2)\nabla[u(x) + \theta h(x)] \cdot \nabla h(x)d\theta.$$

$$\tag{4.21}$$

Integrating both sides in Ω, we obtain that

$$\int_{\Omega} [Q(|\nabla[u+h]|^2) - Q(|\nabla u|^2)]dx$$

$$= -2 \int_{\Omega} \int_0^1 \{\nabla \cdot [q(|\nabla(u+\theta h)|^2)\nabla(u+\theta h)]\}h \, d\theta dx$$

$$= -2 \left(\int_0^1 f(u+\theta h)d\theta, h \right)_{L_2}.$$

Put now $u = u(t)$ and $h = u(t+\Delta t) - u(t)$. Then, it follows that

$$\frac{1}{\Delta t} \int_{\Omega} [Q(|\nabla u(t+\Delta t)]|^2) - Q(|\nabla u(t)|^2)]dx$$

$$= 2 \left(\int_0^1 f(\theta u(t+\Delta t) + (1-\theta)u(t))d\theta, \frac{u(t+\Delta t) - u(t)}{\Delta t} \right)_{L_2}.$$

As $t \mapsto f(u(t)) \in L_2(\Omega)$ is continuous due to (4.4), we conclude that

$$\frac{1}{\Delta t} \int_{\Omega} [Q(|\nabla u(t+\Delta t)]|^2) - Q(|\nabla u(t)|^2)]dx \to 2(f(u(t)), u'(t))_{L_2}.$$

Consider the inner product between Eq. (4.2) and $u'(t)$ to obtain that

$$\|u'(t)\|_{L_2}^2 + (Au(t), u'(t))_{L_2} = (f(u(t)), u'(t))_{L_2}.$$

Since $(Au(t), u'(t))_{L_2} = \frac{1}{2}\frac{d}{dt}\|A^{\frac{1}{2}}u(t)\|_{L_2}^2$ and since (4.10) is the case, it follows that

$$\frac{1}{2}\frac{d}{dt}\int_{\Omega} |\Delta u(t)|^2 dx - \frac{1}{2}\frac{d}{dt}\int_{\Omega} Q(|\nabla u(t)|^2)dx = -\|u'(t)\|_{L_2}^2.$$

This shows that $\frac{d}{dt}\Phi(u(t)) = -\|u'(t)\|_{L_2}^2 \le 0$ for every $0 \le t < \infty$.

Thereby, the function $\Phi(u)$ defined by (4.20) plays the role of Lyapunov function for the solution $u(t)$. We also notice that

if $\frac{d}{dt}\Phi(u(t)) = 0$ at $t = \bar{t}$ then $Au(\bar{t}) = f(u(\bar{t}))$,

$$\text{i.e., } u(\bar{t}) \text{ is a stationary solution of (4.2).} \qquad (4.22)$$

4.4.2 ω-Limit Set

For the solution $u(t)$, consider its ω-limit set

$$\omega(u) = \{\overline{u}; \exists t_n \nearrow \infty \text{ such that } u(t_n) \to \overline{u} \text{ in } L_2(\Omega)\}.$$

As verified by (2.11), $\omega(u)$ is a nonempty set and, if $\overline{u} \in \omega(u)$, then $\overline{u} \in \mathcal{D}(A)$ and there exists a temporal sequence $t_m \nearrow \infty$ such that $Au(t_m) \to A\overline{u}$ weakly in $L_2(\Omega)$ and $u(t_m) \to \overline{u}$ strongly in $L_2(\Omega)$.

Proposition 4.4 *There exists an ω-limit $\overline{u} \in \omega(u)$ which is a stationary solution of* (4.2).

Proof Integrating the equality $-\frac{d}{dt}\Phi(u(t)) = \|u'(t)\|_{L_2}^2$ on $[0, \infty)$, we have

$$\int_0^\infty \|u'(t)\|_{L_2}^2 dt = \Phi(u(0)) - \inf_{0 < t < \infty} \Phi(u(t)) < \infty.$$

For the same reasons as for (3.46), there exists a temporal sequence $t_m \nearrow \infty$ for which it holds that $u'(t_m) \to 0$ in $L_2(\Omega)$. Then, there exist a subsequence $t_{m'}$ of t_m and an ω-limit $\overline{u}$ such that $Au(t_{m'}) \to A\overline{u}$ weakly and $u(t_{m'}) \to \overline{u}$ strongly. Consequently, $A^\theta u(t_{m'}) \to A^\theta \overline{u}$ strongly in $L_2(\Omega)$ for any $0 < \theta < 1$. Due to (4.4), $A^\beta u(t_{m'}) \to A^\beta \overline{u}$ in $L_2(\Omega)$ implies $f(u(t_{m'})) \to f(\overline{u})$ in $L_2(\Omega)$. Therefore, letting $m' \to \infty$ in the equation $u'(t_{m'}) + Au(t_{m'}) = f(u(t_{m'}))$, we obtain that $A\overline{u} = f(\overline{u})$.   □

4.5 Formulation

Let A be the positive definite self-adjoint operator of $L_2(\Omega)$ defined by (4.9). By assumption, our solution $u(t)$ belongs to $\mathcal{C}([0, \infty); \mathcal{D}(A)) \cap \mathcal{C}^1([0, \infty); L_2(\Omega))$. Meanwhile, the Lyapunov function $\Phi(u)$ given by (4.20) is defined on $\overset{\circ}{H}^2(\Omega) = \mathcal{D}(A^{\frac{1}{2}})$. Taking account of these situations, we are naturally led to set the triplet $Z \subset X \subset Z^*$ as

$$Z = \mathcal{D}(A), \quad X = \mathcal{D}(A^{\frac{1}{2}}), \quad Z^* = \mathcal{D}(A^0) = L_2(\Omega). \tag{4.23}$$

Obviously, Z (resp. X) is a Hilbert space equipped with the inner product $(A\cdot, A\cdot)_{L_2}$ (resp. $(A^{\frac{1}{2}}\cdot, A^{\frac{1}{2}}\cdot)_{L_2}$). So, the duality product between Z and Z^* is given by

$$\langle u, f \rangle_{Z \times Z^*} = (Au, f)_{L_2}, \qquad (u, f) \in Z \times Z^*$$

$$\langle u, v \rangle_{Z \times Z^*} = (A^{\frac{1}{2}}u, A^{\frac{1}{2}}v)_{L_2}, \qquad (u, v) \in Z \times X.$$

Due to (4.16), $u(t)$ satisfies, in addition to (2.3), the boundedness (2.4).

In order to verify the differentiability of $\Phi\colon X \to \mathbb{R}$, we have to extend the domain $\mathcal{D}(f)$ of f given by (4.3) to the whole space X. In view of (4.7), we define

$$\mathcal{F}(u) = -\nabla \cdot [q(|\nabla u|^2)\nabla u], \qquad u \in \overset{\circ}{H}{}^2(\Omega). \tag{4.24}$$

Since D_{x_i} ($i = 1, 2$) is a continuous operator from $L_2(\Omega)$ into $H^{-1}(\Omega)$, $\mathcal{F}$ is considered to be a mapping from X into $H^{-1}(\Omega) \subset H^{-2}(\Omega)$.

Proposition 4.5 *The function $\Phi\colon X \to \mathbb{R}$ is continuously differentiable in the sense of (2.5)–(2.6) with the derivative*

$$\dot{\Phi}(u) = u - \mathcal{A}^{-1}\mathcal{F}(u), \qquad u \in X. \tag{4.25}$$

Proof For $u, h \in X$, we have

$$a\|\Delta(u + h)\|_{L_2}^2 - a\|\Delta u\|_{L_2}^2 - 2a(\Delta u, \Delta h) = a\|\Delta h\|_{L_2}^2.$$

Therefore, on account of (4.10),

$$a\|\Delta(u + h)\|_{L_2}^2 - a\|\Delta u\|_{L_2}^2 - 2(u, h)_X = \|h\|_X^2. \tag{4.26}$$

In the meantime, we notice that the formula (4.21) is available for a.e. $x \in \Omega$ to u, h, too. So, we can write

$$Q(|\nabla[u(x) + h(x)]|^2) - Q(|\nabla u(x)|^2) - 2q(|\nabla u(x)|^2)\nabla u(x) \cdot \nabla h(x)$$

$$= 2\int_0^1 [q(|\nabla[u(x) + \theta h(x)]|^2) - q(|\nabla u(x)|^2)]\nabla u(x) \cdot \nabla h(x)d\theta$$

$$+ 2\int_0^1 q(|\nabla[u(x) + \theta\nabla h(x)]|^2)\theta|\nabla h(x)|^2 d\theta.$$

Since

$$|q(|\nabla[u(x) + \theta h(x)]|^2) - q(|\nabla u(x)|^2)| \le C[|\nabla u(x)| + |\nabla h(x)|]|\nabla h(x)|$$

due to (4.7), we obtain the estimate

$$|Q(|\nabla[u(x) + h(x)]|^2) - Q(|\nabla u(x)|^2) - 2q(|\nabla u(x)|^2)\nabla u(x) \cdot \nabla h(x)|$$

$$\le C[|\nabla u(x)| + |\nabla h(x)|]|\nabla h(x)|^2.$$

Integration of this inequality in Ω then yields the estimate

$$\left| \int_{\Omega} \left[Q(|\nabla(u+h)|^2) - Q(|\nabla u|^2) \right] dx - 2(q(|\nabla u|^2)\nabla u, \nabla h)_{L_2} \right|$$

$$\leq C(\|\nabla u\|_{L_3} + \|\nabla h\|_{L_3})\|\nabla h\|_{L_3}^2.$$

Here we observe by using (4.11) that $(q(|\nabla u|^2)\nabla u, \nabla h)_{L_2}$ can be rewritten as

$$(q(|\nabla u|^2)\nabla u, \nabla h)_{L_2} = -\left\langle \nabla \cdot [q(|\nabla u|^2)\nabla u], h \right\rangle_{H^{-1} \times \mathring{H}^1} = \langle \mathcal{F}(u), h \rangle_{H^{-1} \times \mathring{H}^1}$$

$$= \langle \mathcal{F}(u), h \rangle_{H^{-2} \times \mathring{H}^2} = (A^{\frac{1}{2}}\mathcal{A}^{-1}\mathcal{F}(u), A^{\frac{1}{2}}h)_{L_2} = (\mathcal{A}^{-1}\mathcal{F}(u), h)_X.$$

Therefore,

$$\left| \int_{\Omega} \left[Q(|\nabla(u+h)|^2) - Q(|\nabla u|^2) \right] dx - 2(\mathcal{A}^{-1}\mathcal{F}(u), h)_X \right|$$

$$\leq C[\|u\|_X + \|h\|_X]\|h\|_X^2. \tag{4.27}$$

Combining (4.26) and (4.27), we conclude the differentiability of $\Phi(u)$ together with (4.25). Obviously, $u \mapsto \dot{\Phi}(u)$ is continuous from X into itself. □

In addition, we have the following result.

Proposition 4.6 *If* $u \in Z$, *then* $\dot{\Phi}(u) = u - A^{-1}f(u) \in Z$ *and the mapping* $u \mapsto \dot{\Phi}(u)$ *is continuous from* Z *into itself.*

Proof If $u \in \mathcal{D}(A)$, then we obtain by virtue of (4.12) or (4.13) that $\mathcal{F}(u) = f(u) \in L_2(\Omega)$; therefore, $\dot{\Phi}(u) = u - A^{-1}f(u) \in \mathcal{D}(A)$. It is easily verified by (4.4) that $u \mapsto \dot{\Phi}(u)$ is continuous from $\mathcal{D}(A)$ into itself. □

As for an ω-limit in the set of (2.11), we choose the $\overline{u}$ which is a stationary solution to (4.2) whose existence has been verified by Proposition 4.4.

In the subsequent sections, we shall show that the structural assumptions announced in Section 2.2 are all fulfilled by this $\overline{u} \in \omega(u)$.

4.6 Verification of Structural Assumptions

In this section, let us verify the Critical Condition, Lyapunov Function and Angle Condition.

(I) *Critical Condition*. The ω-limit $\overline{u}$ has been chosen so that $\overline{u}$ is a stationary solution to (4.2). The formula observed in Proposition 4.6 then shows $\dot{\Phi}(\overline{u}) = 0$.

(II) *Lyapunov Function.* We know that $\frac{d}{dt}\Phi(u(t)) \leq 0$ for any $0 \leq t < \infty$. According to (4.22), if $\frac{d}{dt}\Phi(u(t)) = 0$ at some time $t = \bar{t}$, then $u(\bar{t})$ is a stationary solution of (4.2) and $u(t) = u(\bar{t})$ for all $t \geq \bar{t}$. Thereby, Theorem 4.1 below is automatically valid. For this reason, it suffices to argue under (2.12).

(III) *Angle Condition.* By the setting of spaces (4.23), we can write

$$-\langle \dot{\Phi}(u(t)), u'(t)\rangle_{Z \times Z^*} = -(A\dot{\Phi}(u(t)), u'(t))_{L_2}$$
$$= \| -Au(t) + f(u(t))\|_{L_2}^2 = \|u'(t)\|_{L_2}^2$$
$$= \|\dot{\Phi}(u(t))\|_Z \|u'(t)\|_{Z^*},$$

which shows that (2.13) is fulfilled with $\delta = 1$.

Hence, if we verify the *Gradient Inequality*, then Theorem 2.1 concludes the following convergence theorem of $u(t)$. However, as its verification requires much more essential considerations on $\Phi(u)$ using the abstract results of Section 2.3, the arguments of verifying (2.14) will be described in the next section.

Theorem 4.1 *Let Ω be a two-dimensional, convex-polygonal or $\mathcal{C}^4$, bounded domain, and let $q(\xi) \geq 0$ be a $\mathcal{C}^3$ function defined for $0 \leq \xi < \infty$. Assume that there exists a global solution $u(t)$ to (4.2) in the function space (4.6) and that $q(\xi)$ satisfies the analyticity (4.8). Then, as $t \to \infty$, $u(t)$ converges to a stationary solution $\bar{u}$ of (4.2) at a rate*

$$\|u(t) - \bar{u}\|_{L_2} \leq (D\theta)^{-1}[\Phi(u(t)) - \Phi(\bar{u})]^{\theta} \qquad \text{for all sufficiently large } t$$

with some exponent $0 < \theta \leq \frac{1}{2}$ and constant D appearing in (2.14).

4.7 Gradient Inequality

Let $\Phi(u)$ be the function given by (4.20) on the space $X = \mathring{H}^2(\Omega)$ and let $\bar{u} \in \omega(u)$ be the ω-limit of $u(t)$ fixed above. We begin by showing the continuous Fréchet differentiability of $\dot{\Phi}: X \to X$. In view of (4.25), it is essential to prove the following proposition.

Proposition 4.7 *The mapping $\mathcal{F}: X \to H^{-1}(\Omega)$ defined by (4.24) is continuously Fréchet differentiable with the derivative*

$$\mathcal{F}'(u)h = -\nabla \cdot \left[2q'(|\nabla u|^2)(\nabla u \cdot \nabla h)\nabla u + q(|\nabla u|^2)\nabla h \right], \qquad h \in X. \qquad (4.28)$$

Proof Since D_{x_i} ($i = 1, 2$) is a bounded operator from $L_2(\Omega)$ into $H^{-1}(\Omega)$, it suffices to investigate the mapping $u \mapsto q(|\nabla u|^2)\nabla u$ from X into $[L_2(\Omega)]^2$.

For $u,\ h \in X$, we have

$$q(|\nabla[u+h]|^2)\nabla[u+h] - q(|\nabla u|^2)\nabla u$$
$$= \{q(|\nabla[u+h]|^2) - q(|\nabla u|^2)\}[\nabla u + \nabla h] + q(|\nabla u|^2)\nabla h.$$

So that,

$$q(|\nabla[u+h]|^2)\nabla[u+h] - q(|\nabla u|^2)\nabla u - 2q'(|\nabla u|^2)(\nabla u \cdot \nabla h)\nabla u$$
$$- q(|\nabla u|^2)\nabla h = \{q(|\nabla[u+h]|^2) - q(|\nabla u|^2) - 2q'(|\nabla u|^2)(\nabla u \cdot \nabla h)\}\nabla u$$
$$+ \{q(|\nabla[u+h]|^2) - q(|\nabla u|^2)\}\nabla h. \qquad (4.29)$$

Here, noting that, for a.e. $x \in \Omega$,

$$q(|\nabla[u(x)+h(x)]|^2) - q(|\nabla u(x)|^2) = \int_0^1 \frac{d}{d\theta}\, q(|\nabla[u(x)+\theta h(x)]|^2)d\theta$$
$$= 2\int_0^1 q'(|\nabla[u(x)+\theta h(x)]|^2)[\nabla u(x) \cdot \nabla h(x) + \theta|\nabla h(x)|^2]d\theta,$$

we can write

$$q(|\nabla[u(x)+h(x)]|^2) - q(|\nabla u(x)|^2) - 2q'(|\nabla u(x)|^2)[\nabla u(x) \cdot \nabla h(x)]$$
$$= 2\int_0^1 \{q'(|\nabla[u(x)+\theta h(x)]|^2) - q'(|\nabla u(x)|^2)\}[\nabla u(x) \cdot \nabla h(x)]d\theta$$
$$+ 2\int_0^1 q'(|\nabla[u(x)+\theta h(x)]|^2)\theta|\nabla h(x)|^2 d\theta.$$

Then, on account of (4.7),

$$\left| q(|\nabla[u(x)+h(x)]|^2) - q(|\nabla u(x)|^2) - 2q'(|\nabla u(x)|^2)[\nabla u(x) \cdot \nabla h(x)] \right|$$
$$\leq C[(1+|\nabla u(x)|^2)|\nabla h(x)|^2 + |\nabla u(x)||\nabla h(x)|^3].$$

Using this estimate, we finally obtain from (4.29) that

$$\left\| q(|\nabla[u+h]|^2)\nabla[u+h] - q(|\nabla u|^2)\nabla u - 2q'(|\nabla u|^2)(\nabla u \cdot \nabla h)\nabla u \right.$$
$$\left. - q(|\nabla u|^2)\nabla h \right\|_{L_2} \leq C \sum_{k=2}^3 \|(1+|\nabla u|)^{5-k}|\nabla h|^k\|_{L_2}.$$

Since

$$\|(1 + |\nabla u|)^{5-k}|\nabla h|^k\|_{L_2} \leq \|(1 + |\nabla u|)\|_{L_{10}}^{5-k}\|\nabla h\|_{L_{10}}^k \leq C(1 + \|u\|_X)^{5-k}\|h\|_X^k,$$

we conclude that $u \mapsto q(|\nabla u|^2|)\nabla u$ is Fréchet differentiable from X into $[L_2(\Omega)]^2$.

As remarked at the beginning of the proof, the differentiability of $u \mapsto \mathcal{F}(u)$ from X into $H^{-1}(\Omega)$ now follows together with (4.28).

Continuity of $u \mapsto \mathcal{F}'(u)$ from X into $\mathcal{L}(X, H^{-1}(\Omega))$ is similarly verified.   □

As an immediate consequence, we conclude that $\dot{\Phi} : X \to X$ is continuously Fréchet differentiable with the derivative

$$[\dot{\Phi}]'(u)h = h - A^{-1}\mathcal{F}'(u)h, \qquad h \in X. \tag{4.30}$$

4.7.1 Verification of (2.18)

Let us put $L = [\dot{\Phi}]'(\overline{u})$ and see that L is a Fredholm operator of X. Indeed, due to (4.30), L is given by $L = I - A^{-1}\mathcal{F}'(\overline{u})$; therefore, we can apply Theorem 1.19 if the following proposition is verified.

Proposition 4.8 *The operator $h \mapsto A^{-1}\mathcal{F}'(\overline{u})h$ is a compact operator from X into itself.*

Proof Since $\mathcal{D}(A) \subset H^3(\Omega)$ by Proposition 4.1, we see that $\nabla \overline{u} \in H^2(\Omega)$. Then, it is verified that $\mathcal{F}'(\overline{u})$ is a bounded operator from X into $L_2(\Omega)$. Therefore, we have $A^{-1}\mathcal{F}'(\overline{u}) = A^{-1}\mathcal{F}'(\overline{u})$. For this reason, it suffices to prove that A^{-1} is a compact operator from $L_2(\Omega)$ into X.

Consider any bounded sequence h_m of X. Then, $u_m = A^{-1}h_m$ is a bounded sequence of $\mathcal{D}(A)$. Since the embedding $\mathcal{D}(A) \subset L_2(\Omega)$ is compact, we can extract a subsequence $u_{m'}$ which is convergent in $L_2(\Omega)$. By the moment inequality $\|A^{\frac{1}{2}}u\|_{L_2} \leq C\|Au\|_{L_2}^{\frac{1}{2}}\|u\|_{L_2}^{\frac{1}{2}}$ for $u \in \mathcal{D}(A)$, it follows that $u_{m'}$ is convergent even in $\mathcal{D}(A^{\frac{1}{2}})$. In view of (4.23), we conclude the desired compactness of A^{-1}.   □

4.7.2 Space Y

Let us now set the space Y. Taking account of the definition (4.3) of $f(u)$, we want to set Y as

$$Y = X \cap \mathcal{C}^1(\overline{\Omega}) = \mathring{H}^2(\Omega) \cap \mathcal{C}^1(\overline{\Omega}) \tag{4.31}$$

with the norm $\| \cdot \|_Y = \| \cdot \|_{H^2} + \| \cdot \|_{\mathcal{C}^1}$. Clearly, Y is a dense subspace of X with continuous embedding, i.e., (2.19) is fulfilled.

It is clear that (2.20) is fulfilled, because of $\overline{u} \in \mathcal{D}(A) \subset Y$.

The condition (2.21) is also verified by similar arguments. Let $h \in L^{-1}(Y)$ for $h \in X$; then, $h = Lh + \mathcal{A}^{-1}\mathcal{F}'(\overline{u})h$ and $Lh \in Y$. But, as noticed above, we have $\mathcal{A}^{-1}\mathcal{F}'(\overline{u}) = A^{-1}\mathcal{F}'(\overline{u})$ on X; therefore, it follows that $\mathcal{A}^{-1}\mathcal{F}'(\overline{u})h \in \mathcal{D}(A) \subset Y$. Hence, h must lie in Y.

Let us verify (2.22) and (2.23). By the definitions (4.3) and (4.24), $\mathcal{F}(u) = f(u)$ for $u \in Y$; hence, $\dot{\Phi}(u) = u - A^{-1}f(u)$. For the same reason as above, it follows that $\dot{\Phi}(u) \in Y$ for $u \in Y$. To verify the differentiability of $\dot{\Phi} : Y \to Y$, it is essential to show analogously to Proposition 4.7 that $u \mapsto f(u)$ is differentiable.

Proposition 4.9 *The mapping $f : Y \to L_2(\Omega)$ is continuously Fréchet differentiable with the derivative*

$$f'(u)h = -\nabla \cdot \left[2q'(|\nabla u|^2)(\nabla u \cdot \nabla h)\nabla u + q(|\nabla u|^2)\nabla h \right], \qquad h \in Y. \qquad (4.32)$$

Proof Since D_{x_i} ($i = 1, 2$) is a bounded operator from $H^1(\Omega)$ into $L_2(\Omega)$, it suffices to show that $u \mapsto q(|\nabla u|^2)\nabla u$ is continuously Fréchet differentiable from Y into $[H^1(\Omega)]^2$. In other words, what we have to do is just to estimate the H^1-norm of the right hand side of (4.29) for $u, h \in Y$ by $\|u\|_Y$ and $\|h\|_Y$. But, it is carried out analogously to the above if we notice the fact that $v, w \in Y$ implies $\nabla v \cdot \nabla w \in H^1(\Omega)$ together with the estimate

$$\|\nabla(\nabla v \cdot \nabla w)\|_{L_2} \le C(\|v\|_{H^2}\|w\|_{\mathcal{C}^1} + \|v\|_{\mathcal{C}^1}\|w\|_{H^2}) \le C\|v\|_Y\|w\|_Y, \quad v, w \in Y,$$

by the definition (4.31). In fact, we finally obtain that

$$\big\| q(|\nabla[u + h]|^2)\nabla[u + h] - q(|\nabla u|^2)\nabla u - 2q'(|\nabla u|^2)(\nabla u \cdot \nabla h)\nabla u$$

$$- q(|\nabla u|^2)\nabla h \big\|_{H^1} \le C \sum_{k=2}^{5}(1 + \|u\|_Y)^{7-k}\|h\|_Y^k,$$

which shows that $u \mapsto q(|\nabla u|^2)\nabla u$ is Fréchet differentiable from Y into $[H^1(\Omega)]^2$.

Consequently, $u \mapsto f(u)$ is Fréchet differentiable from Y into $L_2(\Omega)$ and the formula (4.32) is valid.

Continuity of $u \mapsto f'(u)$ from Y into $\mathcal{L}(Y, L_2(\Omega))$ is similarly proved.   □

Since $A^{-1} : L_2(\Omega) \to \mathcal{D}(A)$ is a bounded operator and since $\mathcal{D}(A) \subset Y$ with continuous embedding, Proposition 4.9 directly yields the continuous Fréchet differentiability of $\dot{\Phi} : Y \to Y$ together with

$$[\dot{\Phi}]'(u)h = h - A^{-1}f'(u)h, \qquad h \in Y.$$

We have thus verified, except (2.35), all other structural assumptions of Section 2.3, i.e., (2.5)–(2.6), (2.16), (2.17), (2.18), (2.19), (2.20), (2.21), (2.22) and (2.23).

4.7.3 Verification of (2.35)

It now remains to verify (2.35). For this purpose, we will utilize, as in Subsection 3.6.3, the method of complexification.

In this subsection, the inner product of $\mathbb{C}^2$ is denoted by

$$(\zeta_1, \zeta_2) \cdot (\zeta_1', \zeta_2') = \zeta_1 \overline{\zeta_1'} + \zeta_2 \overline{\zeta_2'}, \qquad (\zeta_1, \zeta_2), (\zeta_1', \zeta_2') \in \mathbb{C}^2.$$

For simplicity, we will also use the notation

$$(\zeta_1, \zeta_2) \cdot \overline{(\zeta_1', \zeta_2')} = (\zeta_1, \zeta_2) \cdot (\overline{\zeta_1'}, \overline{\zeta_2'}) = \zeta_1 \zeta_1' + \zeta_2 \zeta_2', \quad (\zeta_1, \zeta_2), (\zeta_1', \zeta_2') \in \mathbb{C}^2. \tag{4.33}$$

Let us recall that the critical manifold of $\Phi(u)$ was defined by

$$S = \{u \in Y; \ (I - P)\dot{\Phi}(u) = 0\}$$

(see (2.28)). Here, P is an orthogonal projection from X onto $\mathcal{K}(L)$ which is a finite-dimensional subspace of X due to (2.18), and induces an orthogonal decomposition $X = \mathcal{K}(L) + L(X)$. Due to (2.21), $\mathcal{K}(L)$ is included in Y. On the other hand, due to (2.22), L is a mapping from Y into itself. The decomposition (2.27) then provides that the projection P induces a topological decomposition of the Banach space Y, too, into the form

$$Y = \mathcal{K}(L) + L(Y), \tag{4.34}$$

$$P(Y) = \mathcal{K}(L) \quad \text{and} \quad (I - P)(Y) = L(Y). \tag{4.35}$$

Moreover, L is an isomorphism from $L(Y)$ onto itself. As proved by Proposition 2.2, these facts yield that, in a neighborhood of $\overline{u}$ on S, S is a $\mathcal{C}^1$ manifold having the same dimension as $\mathcal{K}(L)$. More precisely, there exists an open neighborhood $U = U_0 \times U_1$ of $\overline{u}$ in Y, where U_0 (resp. U_1) is an open neighborhood of $P\overline{u}$ (resp. $(I - P)\overline{u}$) in $\mathcal{K}(L)$ (resp. $L(Y)$), such that S is represented in U by

$$S \cap U = \{(u_0, g(u_0)); u_0 \in U_0, \ g : U_0 \to U_1\},$$

g being a $\mathcal{C}^1$ mapping from U_0 into U_1 satisfying $g(P\overline{u}) = (I - P)\overline{u}$.

Let $v_1, v_2, \dots, v_N$ be a basis of $\mathcal{K}(L)$, where $N = \dim \mathcal{K}(L)$, and identify $\mathcal{K}(L)$ with $\mathbb{R}^N$ by the correspondence

$$u_0 = \sum_{k=1}^{N} \xi_k v_k \in \mathcal{K}(L) \quad \longleftrightarrow \quad \boldsymbol{\xi} = (\xi_1, \xi_2, \dots, \xi_N) \in \mathbb{R}^N.$$

Let $P\overline{u} \leftrightarrow \overline{\xi}$ and let U_0 correspond to an open neighborhood Ω of $\overline{\xi}$ in $\mathbb{R}^N$. Our goal is then to verify that

the function $\xi \in \Omega \mapsto \phi(\xi) \equiv \Phi\left(\sum_{k=1}^{N} \xi_k v_k + g(\sum_{k=1}^{N} \xi_k v_k)\right)$

is analytic in a neighborhood of $\overline{\xi}$.   (4.36)

Let us begin by resetting the complex spaces $X_\mathbb{C}$ and $Y_\mathbb{C}$ in view of (4.23) and (4.31) to

$$X_\mathbb{C} = \overset{\circ}{H}{}^2(\Omega; \mathbb{C}) \quad \text{and} \quad Y_\mathbb{C} = \overset{\circ}{H}{}^2(\Omega; \mathbb{C}) \cap \mathcal{C}^1(\overline{\Omega}; \mathbb{C}) \qquad (4.37)$$

respectively. Meanwhile, let $A_\mathbb{C}$ denote a realization of $a\Delta^2$ under the homogeneous Dirichlet boundary conditions of second order in the complex L_2-space $L_2(\Omega; \mathbb{C})$. Actually, $A_\mathbb{C}$ can be determined from the symmetric sesquilinear form $a(u, v) = a \int_\Omega \Delta u \overline{\Delta v} dx$ on $X_\mathbb{C}$. Then, $A_\mathbb{C}$ is a positive definite self-adjoint operator of $L_2(\Omega; \mathbb{C})$ and its domain is, as for A, characterized by Proposition 4.1, but $\overset{\circ}{H}{}^2(\Omega; \mathbb{C})$, $H^3(\Omega; \mathbb{C})$ and $H^4(\Omega; \mathbb{C})$ should be substituted with the corresponding ones, respectively. As $a(\cdot, \cdot)$ is symmetric, (4.10) is still valid but it is rewritten as $\mathcal{D}(A_\mathbb{C}^{\frac{1}{2}}) = X_\mathbb{C}$. Moreover, as seen in Subsection 1.4.2, $A_\mathbb{C}$ enjoys the relation

$$A_\mathbb{C} u = A(\operatorname{Re} u) + i A(\operatorname{Im} u), \qquad u \in \mathcal{D}(A_\mathbb{C}),$$

which means that $A_\mathbb{C}$ is a real operator of $L_2(\Omega; \mathbb{C})$ and that $[A_\mathbb{C}]_{|L_2(\Omega)} = A$. Thereby, identifying A with $A_\mathbb{C}$, we can consider A to be a complex linear operator acting in $L_2(\Omega; \mathbb{C})$.

Next, extend $q(\xi)$ to a complex analytic function for suitable complex variables ζ as follows. Taking account of the assumption (4.8), it is possible to choose a complex convex domain $I_\mathbb{C}$ such that $[0, R] \subset I_\mathbb{C}$ and that

$q(\zeta)$ is an analytic function for $\zeta \in I_\mathbb{C}$.

Furthermore, in view of (4.37), there exists a radius $r > 0$ such that, if $u \in B^{Y_\mathbb{C}}(\overline{u}; r)$, then $\nabla u \cdot \overline{\nabla u} \in I_\mathbb{C}$ for all $x \in \overline{\Omega}$.

Such complexification naturally yields that of $f(u)$. Indeed, we put

$$f_\mathbb{C}(u) = -\nabla \cdot [q(\nabla u \cdot \overline{\nabla u}) \overline{\nabla u}], \qquad u \in B^{Y_\mathbb{C}}(\overline{u}; r).$$

Consequently, $u \mapsto \dot{\Phi}(u)$, which was given by (4.25), can be extended to a complex form as

$$u \mapsto [\dot{\Phi}]_\mathbb{C} u = u - A^{-1} f_\mathbb{C}(u), \qquad u \in B^{Y_\mathbb{C}}(\overline{u}; r).$$

It is then possible to verify that $u \mapsto [\dot{\Phi}]_\mathbb{C}(u)$ maps $B^{Y_\mathbb{C}}(\overline{u}; r)$ into $Y_\mathbb{C}$ and is continuously Fréchet differentiable. Indeed, we have the following proposition.

Proposition 4.10 *The mapping* $u \mapsto f_\mathbb{C}(u)$ *from* $B^{Y_\mathbb{C}}(\overline{u}; r)$ *into* $L_2(\Omega; \mathbb{C})$ *is continuously Fréchet differentiable with the derivative*

$$f_\mathbb{C}'(u)h = -\nabla \cdot [2q'(\nabla u \cdot \overline{\nabla u})(\nabla u \cdot \overline{\nabla h})\overline{\nabla u} + q(\nabla u \cdot \overline{\nabla u})\overline{\nabla h}], \quad h \in Y_\mathbb{C}.$$

Proof If we notice Lemma 4.1 below, then the proof of the proposition is carried out in a way similar to the proof of Proposition 4.9. So the detail will be omitted. □

Lemma 4.1 *It holds that*

$$|q''(\zeta') - q''(\zeta)| + |q'(\zeta') - q'(\zeta)| + |q(\zeta') - q(\zeta)| \le C_q |\zeta' - \zeta|, \quad \zeta', \zeta \in I_\mathbb{C},$$

with some constant $C_q > 0$.

Proof As $I_\mathbb{C}$ is a convex domain of $\mathbb{C}$, the result is proved by arguments similar to the proof of Lemma 3.3. □

By Proposition 4.10, it is easily verified that, if $u \in B^{Y_\mathbb{C}}(\overline{u}; r)$, then $[\dot{\Phi}]_\mathbb{C}(u) \in Y_\mathbb{C}$ and $u \mapsto [\dot{\Phi}]_\mathbb{C}(u)$ is continuously Fréchet differentiable with the derivative

$$[[\dot{\Phi}]_\mathbb{C}]'(u)h = h - A^{-1} f_\mathbb{C}'(u)h, \quad h \in Y_\mathbb{C}.$$

In view of this, we are led to extend the operator L to a complex form by

$$L_\mathbb{C} h = h - A^{-1} f_\mathbb{C}'(\overline{u})h, \quad h \in Y_\mathbb{C}.$$

Since $f_\mathbb{C}'(\overline{u})$ and A^{-1} are real operators, it naturally follows that

$$L_\mathbb{C} h = L(\text{Re}\, h) + iL(\text{Im}\, h), \quad h \in Y_\mathbb{C},$$

which means that $L_\mathbb{C}$ is also a real linear operator on $Y_\mathbb{C}$ whose real part is the L.

Let $h \in \mathcal{K}(L_\mathbb{C})$, i.e., $L_\mathbb{C} h = 0$; it then follows that $L(\text{Re}\, h) = L(\text{Im}\, h) = 0$; therefore, both $\text{Re}\, h$ and $\text{Im}\, h$ belong to $\mathcal{K}(L)$; namely, we have $\mathcal{K}(L_\mathbb{C}) = \mathcal{K}(L) + i\mathcal{K}(L)$. Furthermore, this means that any basis of $\mathcal{K}(L)$ (which consists of real-valued functions) becomes a basis of $\mathcal{K}(L_\mathbb{C})$ in $Y_\mathbb{C}$, too. In particular, we see that $\dim \mathcal{K}(L_\mathbb{C}) = \dim \mathcal{K}(L)$. By the same reasoning, $h \in L_\mathbb{C}(Y_\mathbb{C})$ if and only if both $\text{Re}\, h$ and $\text{Im}\, h$ belong to $L(Y)$, i.e., $L_\mathbb{C}(Y_\mathbb{C}) = L(Y) + iL(Y)$. In particular, $L_\mathbb{C}(Y_\mathbb{C})$ is a closed subspace of $Y_\mathbb{C}$. Then, (4.34) yields that $Y_\mathbb{C}$ is topologically decomposed into

$$\begin{aligned} Y_\mathbb{C} &= Y + iY = [\mathcal{K}(L) + L(Y)] + i[\mathcal{K}(L) + L(Y)] \\ &= [\mathcal{K}(L) + i\mathcal{K}(L)] + [L(Y) + iL(Y)] = \mathcal{K}(L_\mathbb{C}) + L_\mathbb{C}(Y_\mathbb{C}). \end{aligned} \quad (4.38)$$

In view of (4.35), set

$$P_{\mathbb{C}}h = P(\text{Re}\,h) + iP(\text{Im}\,h), \qquad h \in Y_{\mathbb{C}}.$$

Then, $P_{\mathbb{C}}$ becomes a complex bounded linear operator on $Y_{\mathbb{C}}$ and is a projection of the decomposition above, namely, $\mathcal{K}(L_{\mathbb{C}}) = P_{\mathbb{C}}(Y_{\mathbb{C}})$ and $L_{\mathbb{C}}(Y_{\mathbb{C}}) = (I - P_{\mathbb{C}})(Y_{\mathbb{C}})$. Of course, $L_{\mathbb{C}}$ is an isomorphism from $L(Y_{\mathbb{C}})$ onto itself.

Identifying L (resp. P) with $L_{\mathbb{C}}$ (resp. $P_{\mathbb{C}}$), let us consider L (resp. P) to be a complex bounded linear operator on $Y_{\mathbb{C}}$.

We are now ready to define the complex critical manifold of S. In fact, we set

$$S_{\mathbb{C}} = \{u \in B^{Y_{\mathbb{C}}}(\overline{u}; r); \ (I - P)[\dot{\Phi}]_{\mathbb{C}}(u) = 0\}$$

in the neighborhood $B^{Y_{\mathbb{C}}}(\overline{u}; r)$. Then, since $Y_{\mathbb{C}}$ is decomposed into the form (4.38), $[\dot{\Phi}]_{\mathbb{C}}(u)$ is continuously Fréchet differentiable with $\big[[\dot{\Phi}]_{\mathbb{C}}\big]'(\overline{u}) = L$ and since $(I - P)L = L$ is an isomorphism of $L(Y_{\mathbb{C}})$, it is possible to repeat the same arguments as in the proof of Preposition 2.2 to claim that there is an open neighborhood $U = U_0 \times U_1$ of $\overline{u}$ in $Y_{\mathbb{C}}$, where U_0 (resp. U_1) is an open neighborhood of $P\overline{u}$ (resp. $(I - P)\overline{u}$) in $\mathcal{K}(L)$ (resp. $L(Y_{\mathbb{C}})$), such that $S_{\mathbb{C}}$ is given in U by

$$S_{\mathbb{C}} \cap U = \{(u_0, g_{\mathbb{C}}(u_0)); \ u_0 \in U_0, \ g_{\mathbb{C}}: U_0 \to U_1\},$$

where $g_{\mathbb{C}}$ is a mapping from U_0 into U_1 satisfying $g_{\mathbb{C}}(P\overline{u}) = (I - P)\overline{u}$ and is continuously Fréchet differentiable. Hence, $S_{\mathbb{C}}$ is a complex $\mathcal{C}^1$-manifold having the same dimension as $\mathcal{K}(L)$.

As noticed above, the real basis $v_1, v_2, \ldots, v_N$ of the real $\mathcal{K}(L)$ is still a basis of the complex $\mathcal{K}(L)$ in $Y_{\mathbb{C}}$. So, identify $\mathcal{K}(L)$ with $\mathbb{C}^N$ by the correspondence

$$u_0 = \sum_{k=1}^{N} \zeta_k v_k \in \mathcal{K}(L) \qquad \longleftrightarrow \qquad \zeta = (\zeta_1, \zeta_2, \ldots, \zeta_N) \in \mathbb{C}^N.$$

Let U_0 correspond to an open neighborhood $\Omega_{\mathbb{C}}$ of $\overline{\xi}$ in $\mathbb{C}^N$. In $\Omega_{\mathbb{C}}$, we want to consider the function

$$\phi_{\mathbb{C}}(\zeta) = \Phi_{\mathbb{C}}\left(\sum_{k=1}^{N} \zeta_k v_k + g_{\mathbb{C}}\left(\sum_{k=1}^{N} \zeta_k v_k\right)\right), \qquad \zeta \in \Omega_{\mathbb{C}},$$

introducing a complexification $\Phi_{\mathbb{C}}(u)$ of $\Phi(u)$ given by

$$\Phi_{\mathbb{C}}(u) = \frac{1}{2}\int_{\Omega}\left[a(\Delta u)^2 - Q(\nabla u \cdot \overline{\nabla u})\right]dx, \qquad u \in B^{Y_{\mathbb{C}}}(\overline{u}; r),$$

where $Q(\zeta)$ is a primitive of $q(\zeta)$ in $I_{\mathbb{C}}$ determined by $Q(\zeta) = \int_0^{\zeta} q(\zeta')d\zeta'$. Of course, $Q(\zeta)$ coincides with $Q(\xi)$ if $\zeta = \xi$ is a real number in $I_{\mathbb{C}}$. It is then seen that $\phi_{\mathbb{C}}(\zeta)$ is continuously differentiable for each complex variable ζ_k (notice (4.33)). The characterization of analytic functions of several complex variables (see

[Die60, (9.10.1)] or [Hör90, Theorem 2.2.8]) is now available to $\phi_C(\zeta)$ to conclude its analyticity in Ω_C. As a consequence, (4.36) has been verified.

In this way, we see that (2.35) is fulfilled. Theorem 2.2 ultimately provides that there exists an exponent $0 < \theta \leq \frac{1}{2}$ for which it holds that

$$\|\dot{\Phi}(u)\|_Y \geq C|\Phi(u) - \Phi(\overline{u})|^{1-\theta}, \qquad u \in B^Y(\overline{u}; r), \tag{4.39}$$

with some radius $r > 0$ and constant $C > 0$.

4.7.4   Verification of (2.14)

It is now easy to derive from (4.39) the gradient inequality (2.14). Indeed, from (4.12)–(4.13), (4.23) and (4.31), we see that $Z \subset Y$. In addition, Proposition 4.2 provides existence of an exponent $\frac{1}{2} < \alpha < \beta$ such that $\mathcal{D}(A^\alpha) \subset Y$. Then, the moment inequality $\|A^\alpha u\|_{L_2} \leq C\|Au\|_{L_2}^{\alpha-\frac{1}{2}}\|A^{\frac{1}{2}}u\|_{L_2}^{1-\alpha}$ for $u \in \mathcal{D}(A)$ yields $\|u\|_Y \leq C\|u\|_Z^{\alpha-\frac{1}{2}}\|u\|_X^{1-\alpha}$ for $u \in Z$. Therefore, by (4.16), there exists a radius $r''' > 0$ such that $u(t) \in B^X(\overline{u}; r''')$ implies that $u(t) \in B^Y(\overline{u}; r)$, r being as in (4.39). Therefore, as $\|\dot{\Phi}(u)\|_Y \leq C\|\dot{\Phi}(u)\|_Z$, (2.14) is fulfilled.

4.8   Notes and Future Studies

The model equation (4.1) was presented in the paper [JOHGSSO94] for describing the growing process of a crystal surface under molecular beam epitaxy. The function $u(x, t)$ denotes a displacement of surface height from the standard level at $(x, t) \in \Omega \times [0, \infty)$. The model consists of two conflicting effects. One is a surface diffusion of adatoms on the crystal surface whose process is expressed by the biharmonic term $-a\Delta^2 u$ (see [Mul57, EH66]). The other is a roughening which is caused by the Schwoebel effect at each step of the crystal. The roughening process is expressed by the nonlinear advection term $-\nabla \cdot [q(|\nabla u|^2)\nabla u]$ (see [SS66]).

According to [JOHGSSO94], one of possible forms of $q(\xi)$ is that $q(\xi) = \frac{1}{\xi+1}$. In this case, Azizi–Yagi [AY17a, AY17b] proved that for any initial value $u_0 \in \mathcal{D}(A^\beta)$, (4.2) with initial condition $u(0) = u_0$ possesses a unique global solution $u(t)$ lying in (4.6). Afterwards, Azizi–Mola–Yagi [AMY17] showed the asymptotic convergence of $u(t)$ to a stationary solution of (4.2). However, no general sufficient conditions for $q(\xi)$ are known in order that for any $u_0 \in \mathcal{D}(A^\beta)$, (4.2) with initial condition $u(0) = u_0$ possesses a global solution in the function space (4.6). The results in the chapter are therefore generalizations of those of [AMY17].

As for a convergence result for other fourth-order parabolic equations, we quote Rybka–Hoffmann [RH] in which the Cahn–Hilliard equations are studied.

Chapter 5
Chemotaxis Model

This chapter is devoted to considering a chemotaxis model in biological population dynamics. The model equation includes an effect of attraction by chemical substance which is described by an advection equation. Under suitable assumptions, we shall show that the results reviewed in Chap. 2 are available to the model equation.

5.1 Keller–Segel Equations

Consider the Keller–Segel equations

$$
\begin{cases}
\dfrac{\partial u}{\partial t} = a\Delta u - \mu\nabla \cdot [u\nabla\rho] & \text{in } \Omega \times (0, \infty), \\[2mm]
\dfrac{\partial \rho}{\partial t} = b\Delta\rho - d\rho + vu & \text{in } \Omega \times (0, \infty), \\[2mm]
\dfrac{\partial u}{\partial n} = \dfrac{\partial \rho}{\partial n} = 0 & \text{on } \partial\Omega \times (0, \infty),
\end{cases}
\tag{5.1}
$$

in a one-, two- or three-dimensional bounded domain Ω. The unknown function $u = u(x, t)$ denotes a density of bacteria at position $x \in \Omega$ and time $t \in [0, \infty)$ and the unknown function $\rho = \rho(x, t)$ denotes a concentration at $x \in \Omega$ and $t \in [0, \infty)$ of the chemical substance which attracts bacteria and is secreted by themselves. For a survey of the chemotaxis models, we refer the reader to [Yag10, Chap. 12].

© The Author(s), under exclusive license to Springer Nature Singapore Pte Ltd. 2021
A. Yagi, *Abstract Parabolic Evolution Equations and Łojasiewicz–Simon Inequality II*,
SpringerBriefs in Mathematics, https://doi.org/10.1007/978-981-16-2663-0_5

Depending on the dimension of Ω, our assumption on the regularity of Ω is quite different. We assume in each case the following condition:

$$
\begin{cases}
\text{one-dimensional case, } \Omega \text{ is an interval;} \\
\text{two-dimensional case, } \Omega \text{ is convex or of class } \mathcal{C}^2; \\
\text{three-dimensional case, } \Omega \text{ is of class } \mathcal{C}^\infty.
\end{cases}
\tag{5.2}
$$

At boundary point $x \in \partial\Omega$, $n(x)$ denotes the outer normal vector. Therefore, as seen from (5.1), the homogeneous Neumann boundary conditions are imposed on the unknown functions u and ρ.

We begin by rewriting (5.1) into an abstract parabolic evolution equation

$$
U' + \mathcal{A}U = \mathcal{F}(U), \qquad 0 < t < \infty,
\tag{5.3}
$$

in the product underlying space

$$
\mathcal{X} = \left\{ \begin{pmatrix} f \\ \eta \end{pmatrix} ; \ f \in L_2(\Omega) \ \text{ and } \ \eta \in H_N^2(\Omega) \right\},
$$

where $H_N^2(\Omega)$ is a close subspace of $H^2(\Omega)$ consisting of functions satisfying the homogeneous Neumann boundary conditions on $\partial\Omega$.

The linear operator $\mathcal{A} = \begin{pmatrix} A_1 & 0 \\ -\nu & A_2 \end{pmatrix}$ is an operator matrix acting in $\mathcal{X}$, where A_1 (resp. A_2) is a realization of the elliptic operator $-a\Delta + 1$ (resp. $-b\Delta + d$) in $L_2(\Omega)$ under the homogeneous Neumann boundary conditions on $\partial\Omega$. The operators A_i ($i = 1, 2$) are positive definite self-adjoint operators of $L_2(\Omega)$ whose fractional powers have the domains

$$
\begin{cases}
\mathcal{D}(A_i^\theta) = H^{2\theta}(\Omega) & \text{if } 0 \le \theta < \tfrac{3}{4}; \\
\mathcal{D}(A_i^\theta) = H_N^{2\theta}(\Omega) & \text{if } \tfrac{3}{4} < \theta \le 1; \text{ and} \\
\mathcal{D}(A_i^\theta) = \{u \in H_N^2(\Omega); \Delta u \in \mathcal{D}(A_i^{\theta-1})\} & \text{if } 1 < \theta \le 2,
\end{cases}
\tag{5.4}
$$

respectively. Thereby, the space $\mathcal{X}$ has been set as a product space of $\mathcal{D}(A_1^0) = L_2(\Omega)$ and $\mathcal{D}(A_2^1) = H_N^2(\Omega)$. (It is already known that such a setting is quite reasonable when we establish a priori estimates for local solutions of (5.1), see [Yag10, Chapter 12, Subsection 2.1].) As a consequence, the domain of $\mathcal{A}$ is given as

$$
\mathcal{D}(\mathcal{A}) = \left\{ \begin{pmatrix} u \\ \rho \end{pmatrix} ; \ u \in \mathcal{D}(A_1) = H_N^2(\Omega) \ \text{ and } \ \rho \in \mathcal{D}(A_2^2) = \mathcal{H}_{N^2}^4(\Omega) \right\}.
$$

By definition, $\mathcal{H}_{N^2}^4(\Omega)$ coincides with the space $\{\rho \in H_N^2(\Omega); \Delta\rho \in H_N^2(\Omega)\}$ with the norm $\|\rho\|_{\mathcal{H}^4} = \sqrt{\|\rho\|_{H^2}^2 + \|\Delta\rho\|_{H^2}^2}$. Due to (5.2), in the one- or three-dimensional case, we have $\mathcal{H}_{N^2}^4(\Omega) \subset H^4(\Omega)$ with continuous embedding, but in the two-dimensional case, this inclusion may fail. As A_i ($i = 1, 2$) is a real sectorial operator with angle 0, it is the same for A as an operator of $\mathcal{X}$. According to [Yag10, Proposition 12.1], the domains of A^θ are characterized by

$$\mathcal{D}(A^\theta) = \left\{ \begin{pmatrix} u \\ \rho \end{pmatrix}; \ u \in H^{2\theta}(\Omega) \text{ and } \rho \in \mathcal{H}_N^{2(\theta+1)}(\Omega) \right\}, \qquad 0 < \theta < \tfrac{3}{4},$$

(5.5)

$$\mathcal{D}(A^\theta) = \left\{ \begin{pmatrix} u \\ \rho \end{pmatrix}; \ u \in H_N^{2\theta}(\Omega) \text{ and } \rho \in \mathcal{H}_{N^2}^{2(\theta+1)}(\Omega) \right\}, \qquad \tfrac{3}{4} < \theta \le 1.$$

(5.6)

Meanwhile, $\mathcal{F}$ is a nonlinear operator acting in $\mathcal{X}$ given by

$$\mathcal{F}(U) = \begin{pmatrix} -\mu\nabla \cdot [u\nabla\rho] + u \\ 0 \end{pmatrix}, \qquad U = \begin{pmatrix} u \\ \rho \end{pmatrix} \in \mathcal{D}(\mathcal{F}).$$

Here, we can observe the estimate

$$\|\nabla \cdot [u\nabla\rho]\|_{L_2} \le C \begin{cases} \|u\|_{H^{\frac{1}{2}+\varepsilon}} \|\rho\|_{H^2} & \text{(one-dimensional case)}, \\ \|u\|_{H^{1+\varepsilon}} \|\rho\|_{H^2} & \text{(two-dimensional case)}, \\ \|u\|_{H^{\frac{3}{2}+\varepsilon}} \|\rho\|_{H^2} & \text{(three-dimensional case)}, \end{cases}$$

(5.7)

ε being any positive number. On account of this estimate and (5.5)–(5.6), $\mathcal{D}(F)$ can be set so that $\mathcal{D}(\mathcal{F}) \supset \mathcal{D}(A^{\frac{1}{4}+\frac{\varepsilon}{2}})$, $\mathcal{D}(A^{\frac{1}{2}+\frac{\varepsilon}{2}})$ or $\mathcal{D}(A^{\frac{3}{4}+\frac{\varepsilon}{2}})$ is valid for one-, two- or three-dimensional case, respectively.

By the facts (5.5), (5.6) and (5.7), the local existence of solutions to (5.3) can be obtained immediately using Theorem 1.4 with $\beta = 0$ and $\eta = \frac{1}{4} + \frac{\varepsilon}{2}, \frac{1}{2} + \frac{\varepsilon}{2}$ or $\frac{3}{4} + \frac{\varepsilon}{2}$ in the one-, two- or three-dimensional case, respectively. In fact, for any initial value

$$U_0 = {}^t(u_0, \rho_0) \in \mathcal{X} \quad \text{such that } u_0 \ge 0 \text{ and } \rho_0 \ge 0,$$

(5.3) possesses a unique local solution $U(t)$ belonging to the function space

$$0 \le U \in \mathcal{C}((0, T_{U_0}]; \mathcal{D}(A)) \cap \mathcal{C}([0, T_{U_0}]; \mathcal{X}) \cap \mathcal{C}^1((0, T_{U_0}]; \mathcal{X})$$

(5.8)

and satisfying the estimates

$$\|\mathcal{A}U(t)\|_{\mathcal{X}} + \|U'(t)\|_{\mathcal{X}} \le C_{U_0}/t, \qquad 0 < t \le T_{U_0}. \tag{5.9}$$

Here, the time interval T_{U_0} and the constant C_{U_0} are determined by the magnitude of the norm $\|U_0\|_{\mathcal{X}}$ alone.

However, the situation of global existence is quite different from that of local existence. For the one-dimensional case, for every initial value U_0, the local solution in (5.8) can be extended as a global solution on $[0, \infty)$ which is globally bounded (Osaki–Yagi [OY01]). For the two-dimensional case, if the norm $\|u_0\|_{L_1}$ is sufficiently small, then the local solution can be extended as a global solution ([Yag10, (12.22)]); but, in general, the local solution may blow up in a finite time (Herrero–Velázquez [HV96, HV97] and Horstmann–Wang [HW01]). For the three-dimensional case, no sufficient condition is known for the global existence and, even if an initial value U_0 is small, the local solution can blow up in a finite time (Winkler [Wi10]).

In view of the above facts, let us fix a solution $U(t)$ to be studied. First, (5.8) suggests that without loss of generality we can take the initial value U_0 so that

$$0 \le U_0 = {}^t(u_0, \rho_0) \in \mathcal{D}(\mathcal{A}). \tag{5.10}$$

In addition, we assume more strongly that

$$u_0(x) \ge \varepsilon_0 \text{ on } \overline{\Omega} \text{ with some constant } \varepsilon_0 > 0. \tag{5.11}$$

Second, let Eq. (5.3) with initial condition $U(0) = U_0$ possess a global solution $U(t) = {}^t(u(t), \rho(t))$ in the function space

$$0 \le U \in \mathcal{C}([0, \infty); \mathcal{D}(\mathcal{A})) \cap \mathcal{C}^1((0, \infty); \mathcal{X}). \tag{5.12}$$

Third, we assume that $u(t)$ satisfies the global norm boundedness

$$\sup_{0 \le t < \infty} \|u(t)\|_{L_2} < \infty. \tag{5.13}$$

As mentioned, this condition is always satisfied in the one-dimensional case.

Our goal is then to prove that $U(t)$ is asymptotically convergent to a stationary solution of (5.3).

5.2 Global Boundedness of Sobolev Norms for $U(t)$

In applying the general results of Section 2.3 to $U(t)$, we need global boundedness of some Sobolev norms of $U(t)$. Especially, we need that of Sobolev norms for all orders in the three-dimensional case. This section is then devoted to deriving those estimates under (5.2) from the conditions (5.12) and (5.13).

5.2.1 One-Dimensional Case

We want to show that $U(t)$ satisfies

$$\|u(t)\|_{H^2} + \|\rho(t)\|_{H^2} \leq C_1, \qquad 0 \leq \forall t < \infty, \tag{5.14}$$

with some constant C_1. But this estimate has essentially been shown in the proof of [OY01, Proposition 4.1]. So, we omit the proof.

5.2.2 Two-Dimensional Case

We want to show that $U(t)$ satisfies

$$\|u(t)\|_{H^2} + \|\rho(t)\|_{H^2} \leq C_2, \qquad \tau_0 \leq \forall t < \infty, \tag{5.15}$$

with an arbitrarily fixed time $\tau_0 > 0$ and some constant C_2.

First, we see that (5.13) implies that

$$U \in \mathcal{B}([0, \infty); \mathcal{X}). \tag{5.16}$$

Indeed, this was already proved in the proof of [Yag10, Proposition 12.2], Step 3.

Next, due to (5.8), $U(t)$ satisfies

$$\|\mathcal{A}U(t)\|_{\mathcal{X}} \leq C_{U_0}/t, \qquad 0 < t \leq T_{U_0},$$

in a neighborhood of the initial time 0, C_{U_0} and T_{U_0} being determined by the magnitude of norm $\|U_0\|_{\mathcal{X}}$ alone.

Then, resetting the initial time to any other $\tau > 0$, we regard $U(t)$ as a solution of (5.3) for $t \in [\tau, \infty)$ with initial value $U(\tau)$. It must then follow that

$$\|\mathcal{A}U(t)\|_{\mathcal{X}} \leq C_{U(\tau)}/(t - \tau), \qquad \tau < t \leq \tau + T_{U(\tau)},$$

$C_{U(\tau)} > 0$ and $T_{U(\tau)} > 0$ being determined by the norm $\|U(\tau)\|_{\mathcal{X}}$ alone. But, as (5.16) is the case, we see that $C_{U(\tau)}$ and $T_{U(\tau)}$ can be determined uniformly in initial time $\tau \, (> 0)$. Hence, (5.15) is verified.

5.2.3 Three-Dimensional Case

We want to prove that $U(t)$ enjoys infinite temporal–spatial regularities. But, in order to prove these, rather long and careful calculations are required using the three-dimensional Gagliardo–Nirenberg inequality and refined results on abstract parabolic evolution equations. So, here we describe only the outlines of the proof.

We begin by showing that (5.13) yields a global $\mathcal{X}$-norm estimate of $U(t)$.

Proposition 5.1 *The condition (5.13) implies* $\rho \in \mathcal{B}([0, \infty); H_N^2(\Omega))$.

Proof Put $C_u = \sup_{0 \leq t < \infty} \|u(t)\|_{L_2}$. The proof consists of two steps.

Step 1 Regarding $\rho(t)$ as a solution to the equation $\rho' + A_2 \rho = vu(t)$ in $L_2(\Omega)$ for $0 \leq t < \infty$, represent it as $\rho(t) = e^{-tA_2}\rho_0 + v \int_0^t e^{-(t-s)A_2}u(s)ds$ by using the semigroup e^{-tA_2} generated by $-A_2$ on $L_2(\Omega)$. Then, since $A_2 \geq d$, we have the estimate $\|A_2^\theta e^{-tA_2}\|_{\mathcal{L}(L_2)} \leq C_\theta(1+t^{-\theta})e^{-dt}$. Therefore, for any $0 \leq \theta < 1$,

$$\|A_2^\theta \rho(t)\|_{L_2} \leq e^{-dt}\|A_2^\theta \rho_0\|_{L_2} + vC_\theta \int_0^t [1 + (t-s)^{-\theta}]e^{-d(t-s)}\|u(s)\|_{L_2}ds$$

$$\leq e^{-dt}\|A_2^\theta \rho_0\|_{L_2} + vC_\theta C_u \int_0^t (1+s^{-\theta})e^{-ds}ds.$$

In view of (5.4), we obtain that $\rho \in \mathcal{B}([0, \infty); H^s(\Omega))$ for any $s < 2$.

Put $C_{\rho,s} = \sup_{0 \leq t < \infty} \|\rho(t)\|_{H^s}$ for $0 \leq s < 2$.

Step 2 Multiply the first equation of (5.1) by $u(t)$ and integrate the product in Ω. Then, by some calculations,

$$\frac{1}{2}\frac{d}{dt}\int_\Omega u^2 dx + a\int_\Omega |\nabla u|^2 dx = \mu \int_\Omega u\nabla\rho \cdot \nabla u\, dx$$

$$= \frac{\mu}{2}\int_\Omega \nabla\rho \cdot \nabla u^2 dx = -\frac{\mu}{2}\int_\Omega u^2 \Delta\rho\, dx \leq \int_\Omega [u^3 + C|\Delta\rho|^3]dx.$$

Here,

$$\|u\|_{L_3}^3 \leq C\|u\|_{H^1}^{\frac{3}{2}}\|u\|_{L_2}^{\frac{3}{2}} \leq C[\|\nabla u\|_{L_2} + C_u]^{\frac{3}{2}}C_u^{\frac{3}{2}} \leq \frac{a}{2}\|\nabla u\|_{L_2}^2 + CC_u^6,$$

$$\|\Delta\rho\|_{L_3}^3 \leq C\|\Delta\rho\|_{H^1}^{\frac{3}{2}}\|\Delta\rho\|_{L_2}^{\frac{3}{2}} \leq C\|\rho\|_{H^3}^{\frac{3}{2}}\|\rho\|_{H^2}^{\frac{3}{2}} \leq C\|\rho\|_{H^3}^{\frac{3}{2}}\|\rho\|_{H^3}^{\frac{3}{8}}\|\rho\|_{H^{\frac{5}{3}}}^{\frac{9}{8}}$$

$$\leq C\|\rho\|_{H^3}^{\frac{15}{8}}C_{\rho,\frac{5}{3}}^{\frac{9}{8}} \leq \varepsilon\|\rho\|_{H^3}^2 + C_\varepsilon C_{\rho,\frac{5}{3}}^{18},$$

with an arbitrary number $\varepsilon > 0$. Therefore, it follows that

$$\frac{d}{dt}\|u(t)\|_{L_2}^2 + a\|\nabla u(t)\|_{L_2}^2 \leq \varepsilon\|\rho(t)\|_{H^3}^2 + C_\varepsilon[C_u^6 + C_{\rho,\frac{5}{3}}^{18}]. \tag{5.17}$$

Meanwhile, take the inner product of the equation $\rho' + A_2\rho = vu(t)$ and $A_2^{\frac{3}{2}}\rho(t)$ in $L_2(\Omega)$. Then,

$$
\frac{1}{2}\frac{d}{dt}\|A_2\rho(t)\|_{L_2}^2 + \|A_2^{\frac{3}{2}}\rho(t)\|_{L_2}^2 = v(u(t), [-b\Delta + d]A_2\rho(t))_{L_2}
$$

$$
= bv(\nabla u(t), \nabla A_2\rho(t))_{L_2} + dv(u(t), A_2\rho(t))_{L_2}
$$

$$
\leq C[\|\nabla u(t)\|_{L_2} + C_u]\|\rho(t)\|_{H^3} \leq \tfrac{1}{2}\|A_2^{\frac{3}{2}}\rho(t)\|_{L_2}^2 + C[\|\nabla u(t)\|_{L_2}^2 + C_u^2],
$$

because of the continuous embedding $\mathcal{D}(A_2^{\frac{3}{2}}) \subset H^3(\Omega)$ (cf. (5.4)). Therefore,

$$
\frac{d}{dt}\|A_2\rho(t)\|_{L_2}^2 + \|A_2^{\frac{3}{2}}\rho(t)\|_{L_2}^2 \leq C[\|\nabla u(t)\|_{L_2}^2 + C_u^2].
$$

Now, multiply the inequality (5.17) by a parameter $\zeta > 0$ and add its product to the inequality above to obtain that

$$
\frac{d}{dt}[\zeta\|u(t)\|_{L_2}^2 + \|A_2\rho(t)\|_{L_2}^2] + (a\zeta - C)\|\nabla u(t)\|_{L_2}^2
$$

$$
+ (1 - \zeta\varepsilon)\|A_2^{\frac{3}{2}}\rho(t)\|_{L_2}^2 \leq \zeta C_\varepsilon[C_u^6 + C_{\rho,\frac{5}{3}}^{18}] + CC_u^2.
$$

If ζ is fixed sufficiently large and ε is taken sufficiently small so that $a\zeta - C > 0$ and $1 - \zeta\varepsilon > 0$, then we obtain a differential inequality

$$
\frac{d}{dt}[\zeta\|u(t)\|^2 + \|A_2\rho(t)\|_{L_2}^2] + c[\zeta\|u(t)\|_{L_2}^2 + \|A_2\rho(t)\|_{L_2}^2] \leq C[C_u^6 + C_{\rho,\frac{5}{3}}^{18}],
$$

with a suitable constant $c > 0$. Solving this inequality, we conclude the desired uniform boundedness of $\|\rho(t)\|_{H^2}$. $\square$

On the basis of this proposition, we will establish the infinite spatial regularity of $U(t)$. Indeed, it holds true for any integer $m = 0, 1, 2, \ldots$ that

$$
\|u(t)\|_{H^{2(m+1)}} + \|\rho(t)\|_{H^{2(m+2)}} \leq C_{3,m}, \qquad \tau_m \leq \forall t < \infty, \tag{5.18}
$$

with some constant $C_{3,m}$, τ_m being an arbitrarily fixed temporal sequence such that

$$
\tau_{-1} = 0 < \tau_0 < \tau_1 < \tau_2 < \cdots < \tau_m < \tau_{m+1} < \cdots < 1.
$$

In order to prove these regularities, we shall first show infinite temporal regularities for $U(t)$ and then shall obtain the wanted spatial regularities.

5.2.4   Infinite Temporal Regularity

Let us prove that $U(t)$ is infinitely differentiable for $0 < t < \infty$ as a $\mathcal{D}(\mathcal{A})$-valued function by using an induction on $m = 0, 1, 2, \ldots$ for the derivative $U^{(m)}(t)$.

Our induction consists of two assertions. The first one is that $U^{(m)}$ belongs to

$$U^{(m)} \in \mathcal{C}([\tau_m, \infty); \mathcal{D}(\mathcal{A})) \cap \mathcal{C}^1([\tau_m, \infty); \mathcal{X}) \tag{5.19}$$

and $U^{(m)}(t)$ satisfies the uniform estimate

$$\|\mathcal{A}U^{(m)}(t)\|_{\mathcal{X}} + \|U^{(m+1)}(t)\|_{\mathcal{X}} \le C_m, \qquad \tau_m \le \forall t < \infty, \tag{5.20}$$

with some constant C_m. Furthermore, $U^{(m)}(t)$ satisfies the locally uniform Hölder condition

$$\|\mathcal{A}[U^{(m)}(t) - U^{(m)}(s)]\|_{\mathcal{X}} + \|U^{(m+1)}(t) - U^{(m+1)}(s)\|_{\mathcal{X}}$$
$$\le L_m |t - s|^{\sigma_m}, \qquad \tau_m \le \forall s, \forall t < \infty, \ |s - t| \le 1, \tag{5.21}$$

with some exponent $\sigma_m > 0$ and some constant L_m.

The second one is that $U^{(m)}$ is characterized as a solution to the initial value problem

$$\begin{cases} [U^{(m)}]' + \mathcal{A}U^{(m)} = \chi^{(m)}(t), & \tau_{m-1} < t < \infty, \\ U^{(m)}(\tau_{m-1}) = U_m, \end{cases} \tag{5.22}$$

in $\mathcal{X}$ with $U_m = U^{(m)}(\tau_{m-1})$. Here, $\chi(t)$ denotes the function $\mathcal{F}(U(t))$ whose m-th derivative is given by

$$\chi^{(m)}(t) = \begin{pmatrix} -\mu \sum_{i=0}^{m} {}_mC_i \nabla \cdot [u^{(m-i)}(t)\nabla\rho^{(i)}(t)] + u^{(m)}(t) \\ 0 \end{pmatrix}. \tag{5.23}$$

(I) *Case $m = 0$* As noticed from (5.8), $U(t)$ satisfies

$$\|\mathcal{A}U(t)\|_{\mathcal{X}} + \|U'(t)\|_{\mathcal{X}} \le C_{U_0}/t, \qquad 0 < t \le T_{U_0}.$$

Then, resetting the initial time to any other $\tau > 0$, we regard $U(t)$ as a solution of (5.3) for $t \in [\tau, \infty)$ with initial value $U(\tau)$. It then follows that

$$\|\mathcal{A}U(t)\|_{\mathcal{X}} + \|U'(t)\|_{\mathcal{X}} \le C_{U(\tau)}/(t - \tau), \qquad \tau < t \le \tau + T_{U(\tau)},$$

$C_{U(\tau)} > 0$ and $T_{U(\tau)} > 0$ being determined by the norm $\|U(\tau)\|_{\mathcal{X}}$ alone. But, since Proposition 5.1 guarantees global boundedness of $\|U(t)\|_{\mathcal{X}}$, it is deduced that $C_{U(\tau)}$ and $T_{U(\tau)}$ can be determined uniformly in initial time $\tau \, (> 0)$. In this way, $U(t)$ is verified to belong to the space (5.19) and to satisfy (5.20) for $m = 0$.

In order to verify (5.21), we appeal to the maximal regularity [Yag10, (4.17)] for the solutions to semilinear equations. Since $U(\tau) \in \mathcal{D}(\mathcal{A})$ for any initial time $\tau \geq \tau_0$ and since (5.20) is valid, $\mathcal{A}U(t)$ and $U'(t)$ are seen to be locally, uniformly Hölder continuous for $\tau_0 \leq t < \infty$ in such a way that

$$\|\mathcal{A}[U(t) - U(s)]\|_{\mathfrak{X}} + \|U'(t) - U'(s)\|_{\mathfrak{X}} \leq L_0 |t - s|^{\sigma_0},$$

$$\tau_0 \leq \forall s, \; \forall t < \infty, \; |s - t| \leq 1,$$

with some exponent $\sigma_0 > 0$ and some constant L_0.

When $m = 0$, (5.22) is trivial.

(II) *Case $m + 1$* Assume the assertions of induction are proved for all integers i such that $0 \leq i \leq m$.

We divide $\chi^{(m)}(t)$ into

$$\chi^{(m)}(t) = \begin{pmatrix} -\mu \nabla \cdot [u^{(m)}(t) \nabla \rho(t)] \\ 0 \end{pmatrix}$$

$$+ \begin{pmatrix} -\mu \sum_{i=1}^{m} {}_m C_i \nabla \cdot [u^{(m-i)}(t) \nabla \rho^{(i)}(t)] + u^{(m)}(t) \\ 0 \end{pmatrix} \equiv \chi_1(t) + \chi_2(t).$$

Then, since $u \in \mathcal{C}^m([\tau_m, \infty); H_N^2(\Omega))$ and $\rho \in \mathcal{C}^{m+1}([\tau_m, \infty); H_N^2(\Omega))$, $\chi_2(t)$ is observed to be continuously differentiable as an $\mathfrak{X}$-valued function (cf. (5.7)). While, we have formally

$$\chi_1'(t) = \begin{pmatrix} -\mu \nabla \cdot [u^{(m+1)}(t) \nabla \rho(t)] \\ 0 \end{pmatrix}$$

$$+ \begin{pmatrix} -\mu \nabla \cdot [u^{(m)}(t) \nabla \rho'(t)] \\ 0 \end{pmatrix} \equiv \chi_{11}(t) + \chi_{12}(t); \qquad (5.24)$$

but, since $\nabla \cdot [u^{(m+1)}(t) \nabla \rho(t)]$ may not be an L_2-valued function, $\chi_{11}(t)$ cannot be defined as an $\mathfrak{X}$-valued function. Taking account of these situations, we are led to introduce an initial value problem

$$\begin{cases} V' + [\mathcal{A} + \mathcal{B}(t)]V = \chi_{12}(t) + \chi_2(t), & \tau_m < t < \infty, \\ V(\tau_m) = U^{(m+1)}(\tau_m), \end{cases} \qquad (5.25)$$

in $\mathfrak{X}$ for an unknown function $V(t)$. Here, $\mathcal{B}(t)$ is a linear operator given by

$$\mathcal{B}(t) = \begin{pmatrix} \mu \nabla \cdot [\cdot \nabla \rho(t)] & 0 \\ 0 & 0 \end{pmatrix}, \qquad \tau_m \leq t < \infty.$$

Of course, in view of (5.22), $U^{(m+1)}$ is expected to be a solution of (5.25).

We now employ the theory of nonautonomous linear evolution equations. Let us regard (5.25) as a perturbed equation of the form [Yag10, (3.116)]. By using the known inequality

$$\|\nabla \cdot [u\nabla\rho]\|_{L_2} \leq C\|u\|_{H^1}\|\rho\|_{H^3}, \qquad u \in H^1(\Omega), \quad \rho \in H^3(\Omega),$$

and the characterization (5.5), it is verified that the main assumptions (3.117), (3.118) and (3.119) in [Yag10, Chap. 3, Sec. 11] are valid with $\widetilde{\mu} = 1$ and $\widetilde{\nu} = \frac{1}{2}$. Then, [Yag10, Theorem 3.13] provides existence of an evolution operator $V(t, s)$ generated by $\mathcal{A} + \mathcal{B}(t)$ ($\tau_m \leq t < \infty$). Furthermore, since (5.21) implies that $\chi_{12}(t) + \chi_2(t)$ is Hölder continuous as an $\mathcal{X}$-valued function, [Yag10, Theorem 3.14] provides existence of unique solution $V(t)$ to (5.25) in the function space $\mathcal{C}((\tau_m; \infty); \mathcal{D}(\mathcal{A})) \cap \mathcal{C}^1((\tau_m, \infty); \mathcal{X})$ and the expression

$$V(t) = V(t, \tau_m)U^{(m+1)}(\tau_m) + \int_{\tau_m}^t V(t, s)[\chi_{12}(s) + \chi_2(s)]ds, \qquad \tau_m \leq t < \infty.$$

We can then argue in a way similar to the proof of Proposition 3.1 and can conclude that $V(t)$ coincides with $U^{(m+1)}(t)$ for $\tau_m \leq t < \infty$. In this way, (5.19) is proved for $m + 1$. The estimate (5.20) for $m + 1$ is verified by (5.20) of m and by the locally uniform Hölder continuities of the operator $\mathcal{B}(t)$ and of the functions $\chi_{12}(t)$ and $\chi_2(t)$ which are all implied from the conditions (5.21) up to m. The locally uniform Hölder condition (5.21) for $m + 1$ is verified by the maximal regularity [Yag10, (3.130)] for the solution $V(t)$ of (5.25).

As we know that $u^{(m+1)} \in \mathcal{C}((\tau_m, \infty); H_N^2(\Omega))$, (5.24) is now meaningful as an $\mathcal{X}$-valued function. In other words, the formula (5.23) holds true for $m + 1$ as well as the characterization (5.22) for $U^{(m+1)}$.

5.2.5 Infinite Spatial Regularity

On the basis of temporal regularity (5.19) and (5.20), let us prove the wanted spatial infinite regularity (5.18).

We use again an induction on $m = 0, 1, 2, \ldots$. The assertion of induction for m is that the derivatives $u^{(i)}(t)$ and $\rho^{(i)}$ up to m belong to

$$\begin{cases} u^{(i)} \in \mathcal{C}([\tau_m, \infty); H^{2(m+1-i)}(\Omega)), & 0 \leq \forall i \leq m, \\[2mm] \rho^{(i)} \in \mathcal{C}([\tau_m, \infty); H^{2(m+2-i)}(\Omega)), & 0 \leq \forall i \leq m, \end{cases} \qquad (5.26)$$

respectively, and satisfy the estimate

$$
\begin{cases}
\sum_{i=0}^{m} \|u^{(i)}(t)\|_{H^{2(m+1-i)}} \leq D_m, & \tau_m \leq \forall t < \infty, \\
\sum_{i=0}^{m} \|\rho^{(i)}(t)\|_{H^{2(m+2-i)}} \leq D_m, & \tau_m \leq \forall t < \infty,
\end{cases}
\tag{5.27}
$$

with some constant D_m.

(I) *Case* $m = 0$ The assertions are already proved by (5.19) and (5.20).

(II) *Case* $m + 1$ Assume that (5.26) and (5.27) are proved for an integer $m \geq 0$.

The formula (5.22) reads as

$$
\begin{cases}
a\Delta u^{(m)}(t) = u^{(m+1)}(t) + \mu \sum_{i=0}^{m} {}_m C_i \nabla \cdot [u^{(m-i)}(t)\nabla \rho^{(i)}(t)] & \text{in } L_2(\Omega), \\
b\Delta \rho^{(m)}(t) = \rho^{(m+1)}(t) + d\rho^{(m)}(t) - vu^{(m)}(t) & \text{in } H_N^2(\Omega).
\end{cases}
$$

Here, due to (5.19), we have $u^{(m+1)}(t) \in H^2(\Omega)$ for $\tau_{m+1} \leq t < \infty$. Meanwhile, since $H^s(\Omega)$ is a Banach algebra for $s > \frac{3}{2}$, i.e., $\|uv\|_{H^s} \leq C_s \|u\|_{H^s} \|v\|_{H^s}$ for $u, v \in H^s(\Omega)$, it is observed that $u^{(m)}(t)\nabla \rho(t) \in H^2(\Omega)$ and $u^{(m-i)}(t)\nabla \rho^{(i)}(t) \in H^4(\Omega)$ for $1 \leq i \leq m$; therefore, $a\Delta u^{(m)}(t) \in H^1(\Omega)$. The shift property then yields that $u^{(m)}(t) \in H^3(\Omega)$. This in turn implies that $u^{(m)}(t)\nabla \rho(t) \in H^3(\Omega)$ and consequently $a\Delta u^{(m)}(t) \in H^2(\Omega)$. The shift property finally yields that $u^{(m)}(t) \in H^4(\Omega)$ for $\tau_{m+1} \leq t < \infty$.

On the other hand, due to (5.19), we have $\rho^{(m+1)}(t) \in H^4(\Omega)$ for $\tau_{m+1} \leq t < \infty$. Then, since $u^{(m)}(t) \in H^4(\Omega)$, it follows that $b\Delta \rho^{(m)}(t) \in H^4(\Omega)$. The shift property therefore yields that $\rho^{(m)}(t) \in H^6(\Omega)$ for $\tau_{m+1} \leq t < \infty$.

In the next step, we use (5.22) of $m - 1$, which reads as

$$
\begin{cases}
a\Delta u^{(m-1)}(t) = u^{(m)}(t) + \mu \sum_{i=0}^{m-1} {}_{m-1} C_i \nabla \cdot [u^{(m-1-i)}(t)\nabla \rho^{(i)}(t)] & \text{in } L_2(\Omega), \\
b\Delta \rho^{(m-1)}(t) = \rho^{(m)}(t) + d\rho^{(m-1)}(t) - vu^{(m-1)}(t) & \text{in } H_N^2(\Omega).
\end{cases}
$$

As just seen, we have $u^{(m)}(t) \in H^4(\Omega)$ for $\tau_{m+1} \leq t < \infty$. Then, repeating the same arguments as above, we can verify that $u^{(m-1)}(t) \in H^6(\Omega)$ for $\tau_{m+1} \leq t < \infty$. On the other hand, this together with the known regularity $\rho^{(m)}(t) \in H^6(\Omega)$ yields that $b\rho^{(m-1)}(t) \in H^6(\Omega)$ and $\rho^{(m-1)}(t) \in H^8(\Omega)$ for $\tau_{m+1} \leq t < \infty$.

We can in fact repeat these arguments step by step until arriving at the spatial regularity $u(t) \in H^{2(m+2)}(\Omega)$ and $\rho(t) \in H^{2(m+3)}(\Omega)$ for $\tau_{m+1} \leq t < \infty$. Hence, (5.26) has been proved for $m + 1$.

By the same procedure, (5.27) can also be proved for $m + 1$.

5.3 Some Other Properties of $U(t)$

5.3.1 Strict Positivity of $u(t)$

We want to prove that the condition (5.11) implies strict positivity of $u(t)$ for $0 \leq t < \infty$. This property is very important in constructing our Lyapunov function. Indeed, as seen from (5.29) below, the Lyapunov function $\Phi(U)$ includes a function $u \log u$ defined for $0 \leq u < \infty$; it is possible to extend the Lyapunov function for all $u \in L_2(\Omega)$ and $\rho \in H^1(\Omega)$ if we extend $u \log u$ continuously over $-\infty < u < \infty$ by putting $u \log u \equiv 0$ for $-\infty < u < 0$. But, $u \log u$ extended in such a way that it has a singularity at $u = 0$; as a consequence, $\Phi(U)$ is not differentiable in the sense of (2.5)–(2.6). More precisely, we cannot verify the equality (5.28).

Proposition 5.2 *For every $0 < t < \infty$, it is true that $u(t) > 0$ on $\overline{\Omega}$.*

Proof Fix $\tau > 0$ arbitrarily and put $C_\tau = \mu \max_{(x,t) \in \overline{\Omega} \times [0,\tau]} |\Delta \rho(x,t)|$. Here, notice that it follows from $\rho(t) \in \mathcal{H}^4_{N^2}(\Omega)$ that $\Delta \rho(t) \in H^2_N(\Omega) \subset \mathcal{C}(\overline{\Omega})$.

Introduce a cutoff function $H(\xi)$, $-\infty < \xi < \infty$, such that $H(\xi) = \frac{1}{2}\xi^2$ for $-\infty < \xi < 0$ and $H(\xi) = 0$ for $0 \leq \xi < \infty$. And consider the function

$$\psi(t) = \int_\Omega H(u(x,t) - \varepsilon_0 e^{-C_\tau t})dx, \qquad 0 \leq t \leq \tau,$$

ε_0 being the positive constant in (5.11). Then,

$$\psi'(t) = \int_\Omega H'(u - \varepsilon_0 e^{-C_\tau t})[u_t + \varepsilon_0 C_\tau e^{-C_\tau t}]dx$$

$$= \int_\Omega H'(u - \varepsilon_0 e^{-C_\tau t})[a\Delta u - \mu \nabla \cdot (u\nabla\rho) + \varepsilon_0 C_\tau e^{-C_\tau t}]dx.$$

Here, as $\nabla H'(u - \varepsilon_0 e^{-C_\tau t}) = H''(u - \varepsilon_0 e^{-C_\tau t})\nabla u$, we obtain that

$$a\int_\Omega H'(u - \varepsilon_0 e^{-C_\tau t})\Delta u\, dx = -a\int_\Omega \nabla H'(u - \varepsilon_0 e^{-C_\tau t}) \cdot \nabla u\, dx$$

$$= -a\int_\Omega H''(u - \varepsilon_0 e^{-C_\tau t})|\nabla u|^2 dx \leq 0.$$

Meanwhile, we have

$$-\mu\int_\Omega H'(u - \varepsilon_0 e^{-C_\tau t})\nabla \cdot (u\nabla\rho)dx = \mu\int_\Omega \nabla H'(u - \varepsilon_0 e^{-C_\tau t}) \cdot (u\nabla\rho)dx$$

$$= \mu\int_\Omega (u - \varepsilon_0 e^{-C_\tau t})\nabla H'(u - \varepsilon_0 e^{-C_\tau t}) \cdot \nabla\rho\, dx$$

$$+ \mu\int_\Omega \varepsilon_0 e^{-C_\tau t}\nabla H'(u - \varepsilon_0 e^{-C_\tau t}) \cdot \nabla\rho\, dx.$$

Furthermore,

$$\mu \int_\Omega (u - \varepsilon_0 e^{-C_\tau t}) \nabla H'(u - \varepsilon_0 e^{-C_\tau t}) \cdot \nabla \rho \, dx$$

$$= \mu \int_\Omega H'(u - \varepsilon_0 e^{-C_\tau t}) \nabla H'(u - \varepsilon_0 e^{-C_\tau t}) \cdot \nabla \rho \, dx$$

$$= \tfrac{\mu}{2} \int_\Omega \nabla [H'(u - \varepsilon_0 e^{-C_\tau t})^2] \cdot \nabla \rho \, dx = -\tfrac{\mu}{2} \int_\Omega H'(u - \varepsilon_0 e^{-C_\tau t})^2 \Delta \rho \, dx.$$

Similarly,

$$\mu \int_\Omega \varepsilon_0 e^{-C_\tau t} \nabla H'(u - \varepsilon_0 e^{-C_\tau t}) \cdot \nabla \rho \, dx = -\mu \int_\Omega \varepsilon_0 e^{-C_\tau t} H'(u - \varepsilon_0 e^{-C_\tau t}) \Delta \rho \, dx.$$

Therefore, $\psi'(t)$ is estimated by

$$\psi'(t) \leq -\tfrac{1}{2} \int_\Omega H'(u - \varepsilon_0 e^{-C_\tau t})^2 \mu \Delta \rho \, dx$$

$$+ \int_\Omega H'(u - \varepsilon_0 e^{-C_\tau t}) \varepsilon_0 e^{-C_\tau t} [C_\tau - \mu \Delta \rho] dx.$$

Since $H'(\xi)^2 = 2H(\xi)$ and $H'(\xi) \leq 0$, we finally obtain that $\psi'(t) \leq C_\tau \psi(t)$ and that $\psi(t) \leq \psi(0) e^{C_\tau t}$ for any $0 \leq t \leq \tau$. Thus, $\psi(0) = 0$ implies $\psi(t) \equiv 0$ for $0 \leq t \leq \tau$, i.e., $u(t) \geq \varepsilon_0 e^{-C_\tau t}$ for $0 \leq t \leq \tau$.

In these arguments, $\tau > 0$ was arbitrarily fixed.   □

5.3.2 Lyapunov Function

Under (5.10) and (5.11), let $u(t)$, $\rho(t)$ be a global solution of (5.3) belonging to the function space (5.12). Then, as shown in [Yag10, Chapter 12, Section 2], if $u(t) > 0$ for every $0 \leq t < \infty$, then $u(t)$ satisfies the equality

$$\frac{d}{dt} \int_\Omega \left\{ av[u \log u - u] + \frac{b\mu}{2} |\nabla \rho|^2 + \frac{d\mu}{2} \rho^2 - \mu v u \rho \right\} dx$$

$$= -\int_\Omega \left\{ vu |\nabla [a \log u - \mu \rho]|^2 + \mu \left(\frac{\partial \rho}{\partial t} \right)^2 \right\} dx \leq 0 \qquad (5.28)$$

for $0 \leq t < \infty$. As verified by Proposition 5.2, $u(t) > 0$ is the case for $0 \leq t < \infty$; therefore, we have this equality (5.28) for every $0 \leq t < \infty$. This formula means that the functional

$$\Phi(U) = \int_\Omega \left\{ av[u \log u - u] + \frac{b\mu}{2}|\nabla\rho|^2 + \frac{d\mu}{2}\rho^2 - \mu v u \rho \right\} dx, \quad U = \begin{pmatrix} u \\ \rho \end{pmatrix},$$

(5.29)

defined for $0 \le u \in L_2(\Omega)$ and $\rho \in H^1(\Omega)$ becomes a Lyapunov function for the solution $U(t)$. Setting $u \log u = 0$ for $u = 0$, too, we want to consider $u \log u$ to be a continuous function for $0 \le u < \infty$.

We in fact verify the following proposition.

Proposition 5.3 *The value* $\Phi(U(t))$ *decreases monotonously as* $t \nearrow \infty$. *If* $\frac{d}{dt}[\Phi(U(t))] = 0$ *at some time* $t = \bar{t}$, *then* $U(\bar{t})$ *is a stationary solution of* (5.3).

Proof The first assertion is already verified. To verify the second one, let $\frac{d}{dt}[\Phi(U(t))] = 0$ at $t = \bar{t}$ and put $U(\bar{t}) = {}^t(\bar{u}, \bar{\rho})$. Then, in view of Proposition 5.2, it is observed from (5.28) that

$$\nabla[a \log \bar{u} - \mu\bar{\rho}] = 0 \quad \text{and} \quad b\Delta\bar{\rho} - d\bar{\rho} + v\bar{u} = 0 \quad \text{in } \Omega. \tag{5.30}$$

The first equation yields further the equality

$$\nabla \cdot [a\nabla\bar{u} - \mu\bar{u}\nabla\bar{\rho}] = \nabla \cdot \{\bar{u}\nabla[a \log \bar{u} - \mu\bar{\rho}]\} = 0 \quad \text{in } \Omega. \tag{5.31}$$

This together with the second equation of (5.30) shows that $U(\bar{t})$ is a stationary solution of (5.3).   □

5.3.3   ω-Limit of $U(t)$

For the solution $U(t)$, let us define its ω-limit set by

$$\omega(U) = \left\{ \begin{pmatrix} \bar{u} \\ \bar{\rho} \end{pmatrix}; \ \exists t_m \nearrow \infty, \ u(t_m) \to \bar{u} \text{ in } L_2(\Omega) \text{ and } \rho(t_m) \to \bar{\rho} \text{ in } H^1(\Omega) \right\}.$$

Since the closed unit ball $\overline{B}^{H^2}(0; 1)$ of $H^2(\Omega)$ is sequentially, weakly closed in $H^2(\Omega)$ and is relatively compact in $L_2(\Omega)$ or in $H^1(\Omega)$, it is possible by (5.14), (5.15) and (5.18) to assume that $u(t_m) \to \bar{u}$ and $\rho(t_m) \to \bar{\rho}$ weakly in $H^2(\Omega)$ and strongly in $L_2(\Omega)$ and in $H^1(\Omega)$, respectively. As a result, we know that $u(t_m) \to \bar{u}$ and $\rho(t_m) \to \bar{\rho}$ strongly in any Sobolev space $H^s(\Omega)$ such that $0 \le s < 2$. In particular, we see that $u(t_m) \to \bar{u}$ in $\mathcal{C}(\overline{\Omega})$ and $\bar{u} \ge 0$ on $\overline{\Omega}$.

We know that, as $m \to \infty$, $\Phi(U(t_m)) \to \Phi(\overline{U})$, where $\overline{U} = {}^t(\overline{u}, \overline{\rho})$. Therefore, by Proposition 5.3, it follows that

$$\lim_{m \to \infty} \Phi(U(t_m)) = \inf_{0 \le t < \infty} \Phi(U(t)) = \Phi(\overline{U}) \qquad \text{for any } \overline{U} \in \omega(U). \qquad (5.32)$$

The following proposition then shows us existence of a special ω-limit.

Proposition 5.4 *There exists an ω-limit $\overline{U} = {}^t(\overline{u}, \overline{\rho})$ such that $\overline{u}$ satisfies $\overline{u}(x) > 0$ on $\overline{\Omega}$ and that $\overline{u}$, $\overline{\rho}$ satisfy the equations of (5.30). In particular, $\overline{U}$ is a stationary solution of (5.3).*

Proof We have

$$\Phi(U(0)) - \inf_{0 \le t < \infty} \Phi(U(t)) = -\int_0^\infty \frac{d\Phi}{dt}(t)dt.$$

Then, for the same reasons as for (3.46), it is possible to choose a temporal sequence $t_m \nearrow \infty$ such that $\frac{d\Phi}{dt}(t_m) \to 0$. Furthermore, we can choose from t_m a subsequence $t_{m'}$ for which the sequence $U(t_{m'})$ is convergent to an ω-limit $\overline{U} = {}^t(\overline{u}, \overline{\rho})$ of $U(t)$. Then, it follows from (5.28) that, as $t_{m'} \nearrow \infty$,

$$\sqrt{u(t_{m'})}\nabla[a \log u(t_{m'}) - \mu\rho(t_{m'})] \to 0 \qquad \text{in } L_2(\Omega), \qquad (5.33)$$

$$b\Delta\rho(t_{m'}) - d\rho(t_{m'}) + \nu u(t_{m'}) \to 0 \qquad \text{in } L_2(\Omega). \qquad (5.34)$$

Here, let us observe that $\overline{u}(x) > 0$ on $\overline{\Omega}$. Noticing that $\overline{u} \in \mathcal{C}(\overline{\Omega})$, consider the open subset $\Omega_+ = \{x \in \Omega;\ \overline{u}(x) > 0\}$ of Ω. Since $u(t) \ge 0$ and $\|u(t)\|_{L_1} \equiv \|u_0\|_{L_1}$ imply that $\overline{u} \ge 0$ and $\|\overline{u}\|_{L_1} = \|u_0\|_{L_1}$, Ω_+ is a nonempty set. For each $x_0 \in \Omega_+$, let $r_0 > 0$ such that $\overline{B}(x_0; r_0) \subset \Omega_+$. Then, since $u(t_{m'})$ converges to $\overline{u}$ both in $\mathcal{C}(\overline{B}(x_0; r_0))$ and in $H^1(B(x_0; r_0))$, it follows that $\sqrt{u(t_{m'})}\nabla[a \log u(t_{m'}) - \mu\rho(t_{m'})] \to \sqrt{\overline{u}}\nabla[a \log \overline{u} - \mu\overline{\rho}]$ in $L_2(B(x_0; r_0))$. Hence, (5.33) yields the equality $\sqrt{\overline{u}}\nabla[a \log \overline{u} - \mu\overline{\rho}] = 0$ in $B(x_0; r_0)$, i.e., $\nabla[a \log \overline{u} - \mu\overline{\rho}] = 0$. We thus conclude that

$$\nabla[a \log \overline{u} - \mu\overline{\rho}] = 0 \qquad \text{in } \Omega_+.$$

Let Ω'_+ be any connected component of Ω_+. Then, we must have $a \log \overline{u} - \mu\overline{\rho}$ constant, say C', in Ω'_+, namely, $\overline{u}(x) = e^{[C' + \mu\overline{\rho}(x)]/a}$ for any $x \in \Omega'_+$. This means that $\overline{u}(x) > 0$ up to the boundary points $x \in \partial\Omega'_+$. Consequently, Ω_+ must coincide with the whole Ω. We can then repeat the same arguments to conclude that $\overline{u}(x) > 0$ up to the boundary of Ω, i.e., on $\overline{\Omega}$.

Now, it is seen that $\overline{u}$, $\overline{\rho}$ satisfy the first equation of (5.30).

As Δ is a bounded operator from $H^2(\Omega)$ into $L_2(\Omega)$, the weak convergence $\rho(t_{m'}) \to \overline{\rho}$ in $H^2(\Omega)$ implies that of $\Delta\rho(t_{m'}) \to \Delta\overline{\rho}$ in $L_2(\Omega)$. Hence, it follows from (5.34) that $b\Delta\overline{\rho} - d\overline{\rho} + v\overline{u} = 0$ in Ω. Meanwhile, the first equation of (5.30) implies (5.31). Hence, $\overline{U}$ is verified to be a stationary solution to (5.3).   □

Our goal is now to show that, as $t \to \infty$, $U(t)$ converges to the ω-limit $\overline{U}$.

5.4 Formulation

In order to apply the abstract results of Chap. 2, let us here fix a triplet $Z^* \subset X \subset Z$ and define explicitly a Lyapunov function $\Phi(U)$ for $U \in X$.

5.4.1 Zero-Mean L_2-Space

As a basic property of the Keller–Segel equations, the solution $u(t)$ satisfies $\frac{d}{dt}\int_\Omega u(x,t)dx = 0$ for every $0 < t < \infty$, from which it follows that $\int_\Omega u(x,t)dx \equiv \|u_0\|_{L_1}$. It is thereby convenient to switch unknown functions from $u(t)$ to $v(t)$ by the formula $v(t) = u(t) - f$, where f stands for

$$f = |\Omega|^{-1}\|u_0\|_{L_1}.$$

Then, $v(t)$ has a zero-mean for any $0 \leq t < \infty$ and satisfies the equation

$$\frac{\partial v}{\partial t} = a\Delta v - \mu\nabla \cdot ([v + f]\nabla\rho),$$

together with the boundary conditions $\frac{\partial v}{\partial n} = 0$. Of course, all the members of this equation satisfy the zero-mean condition on Ω.

For this reason, introduce the L_2-space of Ω consisting of zero-mean functions:

$$L_{2,m}(\Omega) = \left\{ v \in L_2(\Omega);\ |\Omega|^{-1}\int_\Omega v(x)dx = 0 \right\}.$$

Clearly, $L_{2,m}(\Omega)$ is a closed subspace of $L_2(\Omega)$ of codimension 1 (namely, an orthogonal complement of the space of constant functions) and its orthogonal projection P_m is given by

$$P_m u = u - |\Omega|^{-1}\int_\Omega u(x)dx, \qquad u \in L_2(\Omega). \tag{5.35}$$

Meanwhile, the co-projection onto the space of constant functions is given by

$$Q_m u = (1 - P_m)u = |\Omega|^{-1}\int_\Omega u(x)dx, \qquad u \in L_2(\Omega).$$

As for $v(t)$, (5.12) is now described as

$$v \in \mathcal{C}([0, \infty); H_N^2(\Omega)) \cap \mathcal{C}^1([0, \infty); L_{2,m}(\Omega)). \qquad (5.36)$$

In connection with this switching, we put $V(t) = U(t) - {}^t(f, 0) = {}^t(u(t) - f, \rho(t))$ for $0 \le t < \infty$, $V(t)$ being obviously a solution to the evolution equation

$$V' + \mathcal{A}[V + {}^t(f, 0)] = \mathcal{F}[V + {}^t(f, 0)], \qquad 0 < t < \infty, \qquad (5.37)$$

in the product space of $L_{2,m}(\Omega)$ and $H_N^2(\Omega)$. All the results shown so far for $U(t)$ can of course read for $V(t)$.

5.4.2 Extension of Lyapunov Function

We want to extend the Lyapunov function $\Phi(U)$ defined by (5.29) smoothly on a whole underlying space. In fact, the function is now rewritten as

$$\Phi\left(V + \binom{f}{0}\right) = \int_\Omega \left\{ av[(v + f)\log(v + f) - (v + f)] \right.$$
$$\left. + \frac{b\mu}{2}|\nabla \rho|^2 + \frac{d\mu}{2}\rho^2 - \mu v(v + f)\rho \right\} dx, \qquad V = \binom{v}{\rho},$$

which is defined for $-f \le v \in L_{2,m}(\Omega)$ and $\rho \in H^1(\Omega)$. Let $\overline{U} = {}^t(\overline{u}, \overline{\rho})$ be the ω-limit obtained by Proposition 5.4 and put $\overline{v} = \overline{u} - f$ and $\overline{V} = {}^t(\overline{v}, \overline{\rho})$. As proved, $\overline{u}(x)$ is strictly positive on $\overline{\Omega}$. So, put $\min_{x \in \overline{\Omega}} \overline{u}(x) = \overline{\varepsilon} > 0$, which implies that

$$\min_{x \in \overline{\Omega}} \overline{v}(x) + f \ge \overline{\varepsilon}. \qquad (5.38)$$

In view of this, let us first extend the function $\log \xi$ of $\xi \in \left(\frac{\overline{\varepsilon}}{2}, \infty\right)$ smoothly on the whole real line $(-\infty, \infty)$ by

$$\log^\sim \xi \equiv 0 \quad \text{for } -\infty < \xi \le 0, \qquad \log^\sim \xi = \log \xi \quad \text{for } \frac{\overline{\varepsilon}}{2} \le \xi < \infty,$$

the values $\log^\sim \xi$ for $0 < \xi < \frac{\overline{\varepsilon}}{2}$ being suitably defined. In addition, we put

$$\ell(\xi) = \xi \log^\sim \xi - \xi, \qquad -\infty < \xi < \infty. \qquad (5.39)$$

Then, we can next extend $\Phi(V + {}^t(f, 0))$ as

$$\Phi(V) = \int_\Omega \left\{ av\ell(v + f) \right.$$
$$\left. + \frac{b\mu}{2} |\nabla\rho|^2 + \frac{d\mu}{2} \rho^2 - \mu v(v + f)\rho \right\} dx, \qquad V = \begin{pmatrix} v \\ \rho \end{pmatrix}, \qquad (5.40)$$

which is defined for any $v \in L_{2,m}(\Omega)$ and any $\rho \in H^1(\Omega)$. By definition, if we put $\bar{r} = \frac{\bar{\varepsilon}}{2}$, then it holds that

$$\Phi(V) = \Phi(V + {}^t(f, 0)) \qquad \text{for} \quad v \in B^{\mathcal{C}}(\bar{v}; \bar{r}).$$

5.4.3   Triplet of Spaces

We are in a position to introduce the triplet of spaces $Z \subset X \subset Z^*$.
In view of (5.40), we will set X as

$$X = \left\{ \begin{pmatrix} v \\ \rho \end{pmatrix}; \ v \in L_{2,m}(\Omega) \ \text{and} \ \rho \in H^1(\Omega) \right\}. \qquad (5.41)$$

It is often very convenient to equip $H^1(\Omega)$ with the inner product

$$(\rho, \varphi)_{H^1} = b(\nabla\rho, \nabla\varphi)_{L_2} + d(\rho, \varphi)_{L_2}, \qquad \rho, \varphi \in H^1(\Omega). \qquad (5.42)$$

Then, it holds that

$$(A_2\rho, \varphi)_{L_2} = (\rho, \varphi)_{H^1} \qquad \text{for} \quad \rho \in \mathcal{D}(A_2), \ \varphi \in L_2(\Omega), \qquad (5.43)$$

where A_2 is the realization of $-b\Delta + d$ in $L_2(\Omega)$ introduced in Section 5.1.
Taking account of conditions (2.1), (2.3), (2.4) and (2.8) to be fulfilled by Z and Z^*, we will set Z and Z^* as

$$Z = \left\{ \begin{pmatrix} v \\ \rho \end{pmatrix}; \ v \in H^1_m(\Omega) \ \text{and} \ \rho \in H^2_N(\Omega) \right\}, \qquad (5.44)$$

$$Z^* = \left\{ \begin{pmatrix} w \\ \varphi \end{pmatrix}; \ w \in H^1_m(\Omega)' \ \text{and} \ \varphi \in L_2(\Omega) \right\},$$

respectively. Here, $H_m^1(\Omega)$ is a closed subspace of $H^1(\Omega)$ consisting of functions of $H^1(\Omega)$ of zero-mean. By virtue of the Poincaré inequality, we have $\|v\|_{L_2} \leq C\|\nabla v\|_{L_2}$ for $v \in H_m^1(\Omega)$. Thereby, $H_m^1(\Omega)$ can be equipped with the inner product

$$(v, w)_{H_m^1} = (\nabla v, \nabla w)_{L_2}, \qquad v, \, w \in H_m^1(\Omega). \tag{5.45}$$

Meanwhile, $H_m^1(\Omega)'$ is an adjoint space of $H^1(\Omega)$ such that

$$\text{the spaces } H_m^1(\Omega) \subset L_{2,m}(\Omega) \subset H_m^1(\Omega)' \text{ form a triplet} \tag{5.46}$$

with duality product $\langle v, w \rangle_{H_m^1 \times H_m^1{}'}$. By virtue of (5.4), $H_N^2(\Omega) \subset H^1(\Omega) \subset L_2(\Omega)$ are a triplet with the duality product $\langle \rho, \varphi \rangle_{H_N^2 \times L_2} = (A_2\rho, \varphi)_{L_2}$. In this way, the spaces $Z \subset X \subset Z^*$ are verified to form a triplet whose duality product is given by

$$\left\langle \begin{pmatrix} v \\ \rho \end{pmatrix}, \begin{pmatrix} w \\ \varphi \end{pmatrix} \right\rangle_{Z \times Z^*} = \langle v, w \rangle_{H^1 \times H_m^1{}'} + (A_2\rho, \varphi)_{L_2}. \tag{5.47}$$

It is clear from (5.36) that $V(t)$ belongs to $\mathcal{C}([0, \infty);\ Z) \cap \mathcal{C}^1([0, \infty);\ Z^*)$. Furthermore, as shown by (5.14), (5.15) and (5.18), $V(t)$ satisfies the global boundedness (2.4), i.e.,

$$\|V(t)\|_Z \leq R \qquad \text{for all} \ \ 0 \leq t < \infty. \tag{5.48}$$

Differentiability of $\Phi(V)$ is verified by the following proposition.

Proposition 5.5 *The function* $\Phi : X \to \mathbb{R}$ *is continuously differentiable in the sense of* (2.5)–(2.6) *and its derivative is given by*

$$\dot{\Phi}(V) = \begin{pmatrix} \nu P_m[a\ell'(v + f) - \mu\rho] \\ \mu[\rho - \nu A_2^{-1}(v + f)] \end{pmatrix}, \qquad V = \begin{pmatrix} v \\ \rho \end{pmatrix} \in X. \tag{5.49}$$

Proof For $v, \, h \in L_{2,m}(\Omega)$, we have

$$\int_\Omega [\ell(v + f + h) - \ell(v + f) - \ell'(v + f)h]dx$$

$$= \int_\Omega \left[\int_0^1 \ell'(v + f + \theta h)h d\theta - \ell'(v + f)h\right] dx$$

$$= \int_\Omega \int_0^1 [\ell'(v + \theta h + f) - \ell'(v + f)] h \, d\theta dx.$$

By the definition (5.39), we see that $\sup_{-\infty < \xi < \infty} |\ell''(\xi)| < \infty$. Therefore,

$$\left| \int_\Omega [\ell(v + f + h) - \ell(v + h) - \ell'(v + f)h] dx \right| \leq C \|h\|_{L_{2,m}}^2.$$

This means that the function $v \mapsto \int_\Omega \ell(v + f) dx$ is Fréchet differentiable in $L_{2,m}(\Omega)$ and its derivative at v is given by $h \mapsto \int_\Omega \ell'(v + f)h \, dx = (P_m \ell'(v + f), h)_{L_{2,m}}$.

Meanwhile, for $\rho, \eta \in H^1(\Omega)$, we have

$$\int_\Omega \left[\frac{b}{2} |\nabla(\rho + \eta)|^2 + \frac{d}{2}(\rho + \eta)^2 - \frac{b}{2} |\nabla\rho|^2 - \frac{d}{2}\rho^2 - b\nabla\rho \cdot \nabla\eta - d\rho\eta \right] dx$$
$$= \int_\Omega [\tfrac{b}{2}|\nabla\eta|^2 + \tfrac{d}{2}\eta^2] dx.$$

In view of (5.42),

$$\left| \int_\Omega \left[\frac{b}{2} |\nabla(\rho + \eta)|^2 + \frac{d}{2}(\rho + \eta)^2 - \frac{b}{2}|\nabla\rho|^2 - \frac{d}{2}\rho^2 \right] dx - (\rho, \eta)_{H^1} \right| \leq C\|\eta\|_{H^1}^2,$$

which means that the function $\rho \mapsto \int_\Omega [\frac{b}{2}|\nabla\rho|^2 + \frac{d}{2}\rho^2] dx$ is Fréchet differentiable in $H^1(\Omega)$ and its derivative at ρ is given by $\eta \mapsto (\rho, \eta)_{H^1}$.

Finally, we have

$$\int_\Omega [(v + f + h)(\rho + \eta) - (v + f)\rho - h\rho - (v + f)\eta] dx = \int_\Omega h\eta \, dx,$$

and observe due to (5.42) that

$$\int_\Omega [h\rho + (v + f)\eta] dx = (P_m \rho, h)_{L_{2,m}} + (A_2^{-1}(v + f), \eta)_{H^1}.$$

This shows that the function $V \mapsto \int_\Omega (v + f)\rho \, dx$ is Fréchet differentiable in X and its derivative at V is given by $H \mapsto (P_m \rho, h)_{L_{2,m}} + (A_2^{-1}(v + f), \eta)_{H^1}$, where $H = {}^t(h, \eta)$.

In this way, the Fréchet differentiability of $\Phi(V)$ together with the formula (5.49) has been verified.

The continuity of $V \mapsto \dot{\Phi}(V)$ in X is also easily verified. $\qquad\qquad\square$

Let us next verify that $\Phi(V)$ satisfies the condition (2.8).

Proposition 5.6 *The mapping $V \mapsto \dot{\Phi}(V)$ is continuous from Z into itself.*

Proof Notice that P_m given by (5.35) is a projection from $H^1(\Omega)$ onto $H_m^1(\Omega)$, too. Therefore, if $V = {}^t(v, \rho) \in Z$, then $\ell'(v + f) \in H^1(\Omega)$ and $P_m[a\ell'(v + f) - \mu\rho] \in$

$H_m^1(\Omega)$. Meanwhile, it is clear that $\rho - \nu A_2^{-1}(\nu + f) \in D(A_2)$. Hence, $V \in Z$ implies that $\dot{\Phi}(V) \in Z$.

In order to see the continuity of $V \mapsto \dot{\Phi}(V)$ in Z, it suffices to prove that $\nu \mapsto \ell'(\nu + f)$ is continuous from $H_m^1(\Omega)$ into $H^1(\Omega)$. So, let $\nu, h \in H_m^1(\Omega)$ and let $h \to 0$ in $H_m^1(\Omega)$. Then,

$$\nabla[\ell'(\nu + h + f) - \ell'(\nu + f)]$$
$$= \ell''(\nu + h + f)\nabla h + [\ell''(\nu + h + f) - \ell''(\nu + f)]\nabla \nu.$$

It is clear that $\|\ell''(\nu + h + f)\nabla h\|_{L_2} \leq C\|h\|_{H_m^1} \to 0$. Meanwhile, suppose that the second term $[\ell''(\nu + h + f) - \ell''(\nu + f)]\nabla \nu$ would not converge to 0 in $L_2(\Omega)$. Then, there are a positive constant $\alpha > 0$ and a sequence h_n ($n = 1, 2, 3, \ldots$) such that $h_n \to 0$ in $H_m^1(\Omega)$ for which it must hold that $\|[\ell''(\nu + h_n + f) - \ell''(\nu + f)]\nabla \nu\|_{L_2} \geq \alpha$. But this is a contradiction, because we can extract a subsequence h_{n_k} of h_n such that $h_{n_k}(x)$ converges to 0 almost everywhere in Ω.

Since, as $h \to 0$, $\|\ell'(\nu + h + f) - \ell'(\nu + f)\|_{L_2} \leq C\|h\|_{L_2} \to 0$, it is concluded that $\|\ell'(\nu + h + f) - \ell'(\nu + f)\|_{H^1} \to 0$. $\qquad \square$

For the function $V(t) = U(t) - {}^t(f, 0)$, let $\overline{V}$ be its ω-limit given by $\overline{V} = \overline{U} - {}^t(f, 0)$, where $\overline{U}$ is the ω-limit of $U(t)$ obtained in Proposition 5.4. In the subsequent sections, we shall show that the structural assumptions announced in Section 2.2 are all fulfilled by this $\overline{V} \in \omega(V)$.

5.5 Verification of Structural Assumptions

In this section, let us verify the Critical Condition, Lyapunov Function and Angle Condition, leaving the Gradient Inequality for the next section.

(I) *Critical Condition.* As $\overline{U} = {}^t(\overline{u}, \overline{\rho})$ satisfies (5.30), $(\overline{v}, \overline{\rho})$ satisfies $\nabla[a\log(\overline{v} + f) - \mu\overline{\rho}] = 0$ and $b\Delta\overline{\rho} - d\overline{\rho} + v(\overline{v} + f) = 0$ in Ω. Then, the first equation implies that $a\log(\overline{v} + f) - \mu\overline{\rho}$ is constant in Ω. We see from (5.35) that $P_m[a\log(\overline{v} + f) - \mu\overline{\rho}] = 0$. As $\log(\overline{v} + f) = \ell'(\overline{v} + f)$, we conclude that $P_m[a\ell'(\overline{v} + f) - \mu\overline{\rho}] = 0$.

Meanwhile, as $A_2\overline{\rho} = -b\Delta\overline{\rho} + d\overline{\rho}$, it follows that $-A_2\overline{\rho} + v(\overline{v} + f) = 0$. Hence, $\dot{\Phi}(\overline{V}) = 0$ is verified.

(II) *Lyapunov Function.* Since $H_N^2(\Omega) \subset C(\overline{\Omega}) \subset L_2(\Omega)$, there exists a radius $r' > 0$ such that $B^{L_2}(\overline{v}; r') \cap \overline{B}^{H^2}(0; R) \subset B^C(\overline{v}; \overline{r})$, here $\overline{r}$ is the radius in (5.29) and R is the constant in (5.48). Therefore, if $V(t) \in B^X(\overline{V}; r')$, then $v(t) \in B^C(\overline{v}; \overline{r})$; consequently, $\Phi(V(t)) = \Phi(U(t))$ and $\frac{d}{dt}\Phi(V(t)) = \frac{d}{dt}\Phi(U(t))$.

According to Proposition 5.3, if $\frac{d}{dt}\Phi(U(t)) = 0$ at some time $t = \overline{t}$, then $U(\overline{t})$ is a stationary solution of (5.3); so, the assertion of Theorem 5.1

to be proved is automatically verified. Thereby, it suffices to argue under the condition that $\frac{d}{dt}\Phi(U(t)) < 0$ for every $0 \leq t < \infty$. In view of (5.32), we are therefore allowed to assume the conditions (2.12).

(III) *Angle Condition.* Let r' be the same as above. Then, the condition (2.13) is fulfilled in the ball $B^X(\overline{V}; r')$.

In fact, for $V(t) \in B^X(\overline{V}; r')$ it is observed by (5.37), (5.47) and (5.49) that

$$
\begin{aligned}
&-\langle \dot{\Phi}(V(t)), V'(t)\rangle_{Z\times Z^*} \\
&= -\langle \nu P_m[a \log(v+f) - \mu\rho], \nabla \cdot \{(v+f)\nabla[a \log(v+f) - \mu\rho]\}\rangle_{H_m^1 \times H_m^{1'}} \\
&\quad - \left(A_2\mu[\rho - \nu A_2^{-1}(v+f)], -A_2\rho + v(v+f)\right)_{L_2}.
\end{aligned}
$$

Here, on account of $\nabla \cdot \{(v+f)\nabla[a \log(v+f) - \mu\rho]\} \in L_{2,m}(\Omega)$, it can be written by the property of dual product of (5.46) as

$$
\begin{aligned}
&-\langle \nu P_m[a \log(v+f) - \mu\rho], \nabla \cdot \{(v+f)\nabla[a \log(v+f) - \mu\rho]\}\rangle_{H_m^1 \times H_m^{1'}} \\
&= -(\nu P_m[a \log(v+f) - \mu\rho], \nabla \cdot \{(v+f)\nabla[a \log(v+f) - \mu\rho]\})_{L_{2,m}} \\
&= \nu\|\sqrt{v+f}\,\nabla[a \log(v+f) - \mu\rho]\|_{L_2}^2.
\end{aligned}
$$

On one hand, since $v + f \geq \frac{\bar{\varepsilon}}{2}$ in Ω, we have

$$
\nu\|\sqrt{v+f}\,\nabla[a \log(v+f) - \mu\rho]\|_{L_2}^2 \geq \frac{\nu\bar{\varepsilon}}{2}\|P_m[a \log(v+f) - \mu\rho]\|_{H_m^1}^2, \qquad (5.50)
$$

where we used the definition (5.45) of the norm $H_m^1(\Omega)$.

On the other hand, since $v + f \leq C$ in Ω, we have

$$
\begin{aligned}
&\|\nabla \cdot \{(v+f)\nabla[a \log(v+f) - \mu\rho]\}\|_{H_m^{1'}}^2 \\
&= \sup_{\|w\|_{H_m^1} \leq 1} |\langle w, \nabla \cdot \{(v+f)\nabla[a \log(v+f) - \mu\rho]\}\rangle_{H_m^1 \times H_m^{1'}}|^2 \\
&= \sup_{\|w\|_{H_m^1} \leq 1} |(\nabla w, (v+f)\nabla[a \log(v+f) - \mu\rho])_{L_2}|^2 \\
&\leq \|(v+f)\nabla[a \log(v+f) - \mu\rho]\|_{L_2}^2 \\
&\leq C\|\sqrt{v+f}\,\nabla[a \log(v+f) - \mu\rho]\|_{L_2}^2. \qquad (5.51)
\end{aligned}
$$

In the meantime, it is clear that

$$- \left(A_2 \mu [\rho - \nu A_2^{-1}(v+f)], -A_2\rho + \nu(v+f) \right)_{L_2}$$
$$= \sqrt{\mu} \, \| \rho - \nu A_2^{-1}(v+f) \|_{H_N^2}^2 = \sqrt{\mu} \, \| A_2\rho - \nu(v+f) \|_{L_2}^2 .$$

This together with (5.50) and (5.51) then yields that

$$- \langle \dot{\Phi}(V(t)), V'(t) \rangle_{Z \times Z^*} \geq \delta \| \dot{\Phi}(V(t)) \|_Z \| V'(t) \|_{Z^*}, \qquad V(t) \in B^X(\overline{V}; r'),$$

with some constant $\delta > 0$.

Hence, if we verify the *Gradient Inequality*, then by Theorem 2.1 we can conclude the following asymptotic convergence of $V(t)$. However, as its verification using the abstract results of Section 2.3 requires much more essential considerations on $\Phi(V)$, the full proof will be described in the next section.

Theorem 5.1 *Let Ω be a one-, two- or three-dimensional bounded domain with the regularity (5.2). For an initial value $^t(u_0, \rho_0)$ satisfying (5.10) and (5.11), assume that there exists a global solution $U(t)$ to (5.3) in the function space (5.12) and that (5.13) is satisfied. Put $V(t) = U(t) - {}^t(f, 0)$, where $f = |\Omega|^{-1} \|u_0\|_{L_1}$. Then, as $t \to \infty$, $V(t)$ converges to a stationary solution $\overline{V}$ of (5.37) at a rate*

$$\| V(t) - \overline{V} \|_{Z^*} \leq (D\delta\theta)^{-1} [\Phi(V(t)) - \Phi(\overline{V})]^\theta \qquad \textit{for all sufficiently large } t$$

$$(5.52)$$

with the exponent $0 < \theta \leq \frac{1}{2}$, the constant $\delta > 0$ and the constant $D > 0$ appearing in (2.14).

5.6 Gradient Inequality for $\Phi(V)$

We begin by verifying the differentiability of $\dot{\Phi}(V)$.

Proposition 5.7 *The mapping $V \mapsto \dot{\Phi}(V)$ in X is Gâteaux differentiable at $\overline{V}$ and the derivative for direction H is given by*

$$[\dot{\Phi}]'(\overline{V})H = \begin{pmatrix} \nu P_m \left[\frac{ah}{\overline{v}+f} - \mu\eta \right] \\ \mu[\eta - \nu A_2^{-1}h] \end{pmatrix}, \qquad H = \begin{pmatrix} h \\ \eta \end{pmatrix} \in X, \qquad (5.53)$$

$[\dot{\Phi}]'(\overline{V})$ *being a bounded linear operator from X into itself.*

Proof As a matter of fact, it suffices to verify that the mapping $v \mapsto \ell'(v + f)$ from $L_{2,m}(\Omega)$ into $L_2(\Omega)$ is Gâteaux differentiable with the derivative $h \mapsto \ell''(v + f)h$ $h \in L_{2,m}(\Omega)$ at $v \in v \in L_{2,m}(\Omega)$. But this is an immediate consequence of Theorem 1.18.

By the definition (5.39), $\ell''(\xi)$ coincides with $\frac{1}{\xi}$ for $\frac{\overline{\varepsilon}}{2} < \xi < \infty$. But, as (5.38) is known, we have $\ell''(\overline{v} + f) = \frac{1}{\overline{v}+f}$.

Finally, it is clear that the mapping given by (5.53) is a bounded linear operator from X into itself.   □

5.6.1   Verification of (2.18)

Put $L = [\dot{\Phi}]'(\overline{V})$, which is as shown above, a bounded linear operator of X. This subsection is devoted to seeing that L is a Fredholm operator of X.

In fact, L is written in the form

$$LH = \begin{pmatrix} av P_m(\frac{h}{\overline{v}+f}) \\ \mu\eta \end{pmatrix} - \mu v \begin{pmatrix} P_m\eta \\ A_2^{-1}h \end{pmatrix}, \qquad H = \begin{pmatrix} h \\ \eta \end{pmatrix} \in X.$$

Here we notice the following fact.

Lemma 5.1 *The mapping $h \mapsto P_m(\frac{h}{\overline{v}+f})$ is a linear isomorphism of $L_{2,m}(\Omega)$.*

Proof of Lemma Put $\overline{a}(x) = \frac{1}{\overline{v}(x)+f}$. Then, since $(\|\overline{v}\|_{\mathcal{C}} + f)^{-1} \leq \overline{a}(x) \leq \overline{\varepsilon}^{-1}$ on $\overline{\Omega}$ due to (5.38), it is directly verified for h, $g \in L_{2,m}(\Omega)$ that $P_m[\overline{a}(x)h] = g$ if and only if

$$h = \overline{a}(x)^{-1}g - \left[\frac{\int_\Omega \overline{a}(y)^{-1}g(y)dy}{\int_\Omega \overline{a}(y)^{-1}dy}\right]\overline{a}(x)^{-1}.$$

This means that $h \mapsto P_m[\overline{a}(x)h]$ admits a bounded inverse on $L_{2,m}(\Omega)$.   □

This lemma provides that $H \mapsto {}^t(av P_m(\frac{h}{\overline{v}+f}), \mu\eta)$ is a linear isomorphism of X. In the meantime, $H \mapsto {}^t(P_m\eta, A_2^{-1}h)$ is a compact operator of X. By Corollary 1.4, we then conclude that L is a Fredholm operator of X with index 0, i.e., dim $\mathcal{K}(L) =$ codim $\mathcal{R}(L)$ is finite.

5.6.2   Space Y

Here we set the space Y as

$$Y = \left\{\begin{pmatrix} v \\ \rho \end{pmatrix}; \ v \in \mathcal{C}_m(\overline{\Omega}) \quad \text{and} \quad \rho \in \mathcal{D}(A_2)\right\}, \tag{5.54}$$

where $\mathcal{C}_m(\overline{\Omega})$ is a closed subspace of $\mathcal{C}(\overline{\Omega})$ consisting of zero-mean continuous functions on $\overline{\Omega}$ and where $\mathcal{D}(A_2)$ is given by (5.4). We notice that the projection P_m introduced in (5.35) clearly maps $\mathcal{C}(\overline{\Omega})$ onto $\mathcal{C}_m(\overline{\Omega})$.

Proposition 5.8 *The mapping $V \mapsto \dot{\Phi}(V)$ is continuous from Y into itself and is continuously Fréchet differentiable in Y with the derivative*

$$[\dot{\Phi}]'(V)H = \begin{pmatrix} \nu P_m[a\ell''(\nu+f)h - \mu\eta] \\ \mu[\eta - \nu A_2^{-1}h] \end{pmatrix}, \quad V = \begin{pmatrix} \nu \\ \rho \end{pmatrix}, H = \begin{pmatrix} h \\ \eta \end{pmatrix} \in Y.$$

Proof As Y is defined by (5.54), it suffices as before to verify that the mapping $v \mapsto \ell'(v + f)$ is continuously Fréchet differentiable from $\mathcal{C}_m(\overline{\Omega})$ into $\mathcal{C}(\overline{\Omega})$. But this is directly verified by Theorem 1.15.

It is also easily verified that $V \mapsto [\dot{\Phi}]'(V)$ is continuous from Y into $\mathcal{L}(Y, Y)$.   □

Let us next verify (2.21), namely, if $LH \in Y$, then $H \in Y$. Indeed, $LH \in Y$ means that $aP_m[\frac{h}{\bar{v}+f}] - \mu P_m\eta \in \mathcal{C}_m(\overline{\Omega})$ and $\eta - \nu A_2^{-1}h \in \mathcal{D}(A_2)$. Then, it is clear that $\eta \in \mathcal{D}(A_2)$. Meanwhile, since $\eta \in \mathcal{D}(A_2) = H_N^2(\Omega) \subset \mathcal{C}(\overline{\Omega})$, it follows that $P_m\eta \in \mathcal{C}_m(\overline{\Omega})$; hence, h is also a continuous function of $\overline{\Omega}$.

We have thus verified, except (2.35), all other structural assumptions of Section 2.3, i.e., (2.5)–(2.6), (2.16), (2.17), (2.18), (2.19), (2.20), (2.21), (2.22) and (2.23).

5.6.3   Verification of (2.35)

Before verifying (2.35), let us recall that the critical manifold of $\Phi(V)$ was defined by

$$S = \{V \in Y; \ (I - P)\dot{\Phi}(V) = 0\}.$$

(see (2.28)). Here, P is an orthogonal projection from X onto $\mathcal{K}(L)$ which is a finite-dimensional subspace of X due to (2.18), and induces an orthogonal decomposition $X = \mathcal{K}(L) + L(X)$. Due to (2.21), $\mathcal{K}(L)$ is included in Y. On the other hand, due to (2.22), L is a mapping from Y into itself. The decomposition (2.27) then provides that the projection P also induces a topological decomposition of the Banach space Y into the form

$$Y = \mathcal{K}(L) + L(Y), \tag{5.55}$$

$$PY = \mathcal{K}(L) \quad \text{and} \quad (I - P)Y = L(Y). \tag{5.56}$$

Moreover, L is an isomorphism from $L(Y)$ onto itself. As proved by Proposition 2.2, these facts yield that, in a neighborhood of $\overline{V}$ on S, S is a $\mathcal{C}^1$ manifold having the same dimension as $\mathcal{K}(L)$. More precisely, there exists an open neighborhood

$U = U_0 \times U_1$ of $\overline{V}$ in Y, where U_0 (resp. U_1) is an open neighborhood of $P\overline{V}$ (resp. $(I - P)\overline{V}$) in $\mathcal{K}(L)$ (resp. $L(Y)$), such that S is represented in U by

$$S \cap U = \{(V_0, g(V_0)); \ V_0 \in U_0, \ g : U_0 \to U_1\},$$

g being a $\mathcal{C}^1$ mapping from U_0 into U_1 satisfying $g(P\overline{V}) = (I - P)\overline{V}$.

Let $V_1^0, \ldots, V_N^0$ be a basis of $\mathcal{K}(L)$, where $N = \dim \mathcal{K}(L)$, and identify $\mathcal{K}(L)$ with the Euclidean space $\mathbb{R}^N$ by the correspondence

$$V_0 = \textstyle\sum_{k=1}^N \xi_k V_k^0 \in \mathcal{K}(L) \qquad \longleftrightarrow \qquad \boldsymbol{\xi} = (\xi_1, \ldots, \xi_N) \in \mathbb{R}^N.$$

Let $P\overline{V} \leftrightarrow \overline{\boldsymbol{\xi}}$ and let U_0 correspond to an open neighborhood $\boldsymbol{\Omega}$ of $\overline{\boldsymbol{\xi}}$ in $\mathbb{R}^N$. Our goal is then to verify that

the function $\boldsymbol{\xi} \in \boldsymbol{\Omega} \mapsto \phi(\boldsymbol{\xi}) \equiv \Phi\left(\sum_{k=1}^N \xi_k V_k^0 + g(\sum_{k=1}^N \xi_k V_k^0) \right)$

is analytic in a neighborhood of $\overline{\boldsymbol{\xi}}$.   (5.57)

As we argued in Subsection 3.6.3, we will employ the method of complexification.

The space X (resp. Y) defined by (5.41) (resp. (5.54)) is extended to a complex form as $X_\mathbb{C} = X + iX$ (resp. $Y_\mathbb{C} = Y + iY$). Meanwhile, let $A_{2,\mathbb{C}}$ be the realization of $-b\Delta + d$ under the homogeneous Neumann boundary conditions in the complex space $L_2(\Omega; \mathbb{C})$. Then, $A_{2,\mathbb{C}}$ is a positive definite self-adjoint operator of $L_2(\Omega; \mathbb{C})$ with the domain $H_N^2(\Omega; \mathbb{C})$ and enjoys the property

$$A_{2,\mathbb{C}}\rho = A_2(\operatorname{Re}\rho) + i A_2(\operatorname{Im}\rho), \qquad \rho \in \mathcal{D}(A_{2,\mathbb{C}}),$$

which means that $A_{2,\mathbb{C}}$ is a real operator. Thereby, identifying A_2 with $A_{2,\mathbb{C}}$, we consider A_2 to be a complex self-adjoint operator of $L_2(\Omega; \mathbb{C})$.

The logarithmic function can be extended in the usual way to an analytic function in the complex domain $\mathbb{C} - (-\infty, \frac{\overline{\varepsilon}}{2}]$, where $\overline{\varepsilon}$ is the positive constant in (5.38). So, $\log(v + f)$ is defined for any $v \in \mathcal{C}_m(\overline{\Omega}; \mathbb{C})$ such that $\|v - \overline{v}\|_\mathcal{C} < \overline{r}$, where $\overline{r} = \frac{\overline{\varepsilon}}{2}$. Therefore, the operator $\dot{\Phi}(V)$ described by (5.49) can be extended as

$$[\dot{\Phi}]_\mathbb{C}(V) = \begin{pmatrix} v P_m[a\log(v+f) - \mu\rho] \\ \mu[\rho - vA_2^{-1}(v+f)] \end{pmatrix}, \qquad V = \begin{pmatrix} v \\ \rho \end{pmatrix} \in B^{Y_\mathbb{C}}(\overline{V}; \overline{r}).$$

Then, by the same proof as for Proposition 5.8, $[\dot{\Phi}]_\mathbb{C}(V)$ is seen to be continuously Fréchet differentiable for $V \in B^{Y_\mathbb{C}}(\overline{V}; \overline{r})$ with the derivative

$$[[\dot{\Phi}]_\mathbb{C}]'(V)H = \begin{pmatrix} v P_m\left[\frac{ah}{v+f} - \mu\eta \right] \\ \mu[\eta - vA_2^{-1}h] \end{pmatrix}, \qquad H = \begin{pmatrix} h \\ \eta \end{pmatrix} \in Y_\mathbb{C}.$$

In view of (5.53), we are naturally led to introduce for the operator L its complex form by $L_\mathbb{C}H = \big[[\dot{\Phi}]_\mathbb{C}\big]'(\overline{V})$. As $L_\mathbb{C}$ enjoys the property $L_\mathbb{C}H = L(\operatorname{Re}H) + iL(\operatorname{Im}H)$, $H \in Y_\mathbb{C})$, $L_\mathbb{C}$ is a real linear operator on $Y_\mathbb{C}$. Therefore, if $L_\mathbb{C}H = 0$, then both $\operatorname{Re}H$ and $\operatorname{Im}H$ belong to $\mathcal{K}(L)$, which means that any real basis of $\mathcal{K}(L)$ becomes a basis of $\mathcal{K}(L_\mathbb{C})$. Furthermore, (5.55) yields that $Y_\mathbb{C}$ can be topologically decomposed into

$$Y_\mathbb{C} = Y + iY = [\mathcal{K}(L) + L(Y)] + i[\mathcal{K}(L) + L(Y)]$$

$$= [\mathcal{K}(L) + i\mathcal{K}(L)] + [L(Y) + iL(Y)] = \mathcal{K}(L_\mathbb{C}) + L_\mathbb{C}(Y_\mathbb{C}).$$

Meanwhile, in view of (5.56), we set $P_\mathbb{C}H = P(\operatorname{Re}H) + iP(\operatorname{Im}H)$. Then, $P_\mathbb{C}$ is a real bounded linear operator of $Y_\mathbb{C}$ and is a projection of this topological direct decomposition, namely, $\mathcal{K}(L_\mathbb{C}) = P_\mathbb{C}Y_\mathbb{C}$ and $L_\mathbb{C}(Y_\mathbb{C}) = (I - P_\mathbb{C})Y_\mathbb{C}$. Of course, $L_\mathbb{C}$ is an isomorphism from $L(Y_\mathbb{C})$ onto itself.

Identifying L (resp. P) with $L_\mathbb{C}$ (resp. $P_\mathbb{C}$), let us consider L (resp. P) to be a complex bounded linear operator on $Y_\mathbb{C}$.

It is now natural to define the complex form of the manifold S as

$$S_\mathbb{C} = \{V \in B^{Y_\mathbb{C}}(\overline{V}; \overline{r}); \ (I - P)[\dot{\Phi}]_\mathbb{C}(V) = 0\}$$

in the neighborhood $B^{Y_\mathbb{C}}(\overline{V}; \overline{r})$ of $\overline{V}$. The implicit function theorem is again available to the equation $(I - P)[\dot{\Phi}]_\mathbb{C}(V) = 0$. We can then claim that, in some neighborhood $U = U_0 \times U_1$ of $\overline{V}$ in $Y_\mathbb{C}$, where U_0 (resp. U_1) is a neighborhood of $P\overline{V}$ (resp. $(I - P)\overline{V}$) in $\mathcal{K}(L)$ (resp. $L(Y_\mathbb{C})$), $S_\mathbb{C}$ is represented as

$$S_\mathbb{C} \cap U = \{(V^0, g_\mathbb{C}(V^0)); \ V^0 \in U_0, \ g_\mathbb{C}: U_0 \to U_1\},$$

$g_\mathbb{C}$ being a $\mathcal{C}^1$ mapping from U_0 into U_1 such that $g_\mathbb{C}(P\overline{V}) = (I - P)\overline{V}$. Hence, $S_\mathbb{C}$ is a complex $\mathcal{C}^1$ manifold of dimension $\dim \mathcal{K}(L)$.

As noticed above, the real basis $V_1^0, V_2^0, \dots, V_N^0$ of the real $\mathcal{K}(L)$ is still a basis of the complex $\mathcal{K}(L)$ in $Y_\mathbb{C}$. So, identify $\mathcal{K}(L)$ with $\mathbb{C}^N$ by the correspondence

$$V_0 = \sum_{k=1}^{N} \zeta_k V_k^0 \in \mathcal{K}(T) \quad \longleftrightarrow \quad \zeta = (\zeta_1, \dots, \zeta_N) \in \mathbb{C}^N.$$

Let U_0 correspond to an open neighborhood $\mathbf{\Omega}_\mathbb{C}$ of $\overline{\xi}$ in $\mathbb{C}^N$. In $\mathbf{\Omega}_\mathbb{C}$, we want to consider the function

$$\phi_\mathbb{C}(\zeta) = \Phi_\mathbb{C}\left(\sum_{k=1}^{N} \zeta_k V_k^0 + g_\mathbb{C}\left(\sum_{k=1}^{N} \zeta_k V_k^0\right)\right), \qquad \zeta \in \mathbf{\Omega}_\mathbb{C},$$

introducing a complexification $\Phi_{\mathbb{C}}(u)$ of $\Phi(u)$ given by

$$\Phi_{\mathbb{C}}(V) = \int_{\Omega} \Big\{ av[(v+f)\log(v+f) - (v+f)]$$

$$+ \frac{b\mu}{2} \sum_i \left(\frac{\partial \rho}{\partial x_i}\right)^2 + \frac{d\mu}{2}\rho^2 - \mu v(v+f)\rho \Big\} dx, \qquad V = \begin{pmatrix} v \\ \rho \end{pmatrix} \in B^{Y_{\mathbb{C}}}(\overline{V}; \overline{r}).$$

It is then seen that $\phi_{\mathbb{C}}(\zeta)$ is continuously differentiable for each complex variable ζ_k. The characterization of analytic functions of several complex variables (see [Die60, (9.10.1)] or [Hör90, Theorem 2.2.8]) is now available to $\phi_{\mathbb{C}}(\zeta)$ to conclude its analyticity in $\Omega_{\mathbb{C}}$. As a consequence, (5.57) has been proved.

In this way, we know that the assumption (2.35) is fulfilled. Theorem 2.2 ultimately provides that there exists an exponent $0 < \theta \le \frac{1}{2}$ for which it holds that

$$\|\dot{\Phi}(V)\|_Y \ge C|\Phi(V) - \Phi(\overline{V})|^{1-\theta}, \qquad V \in B^Y(\overline{V}; r), \tag{5.58}$$

with some radius $r > 0$ and constant $C > 0$.

5.6.4 Verification of (2.14)

Finally, on the basis of (5.58), let us verify validity of the condition (2.14). It is in fact possible to use the techniques suggested in Subsection 2.3.5.

(I) *One-Dimensional Case.* From (5.44) and (5.54), we see that $Z \subset Y$. Moreover, in view of $H_m^1(\Omega) \subset \mathcal{C}(\overline{\Omega})$, we have $\|v\|_{\mathcal{C}} \le C\|v\|_{H^2}^{\frac{1}{2}}\|v\|_{L_2}^{\frac{1}{2}}$, $v \in H_m^2(\Omega)$. Therefore, there exists a radius $r''' > 0$ such that $\overline{B}^{H^2}(0; C_1) \cap B^{L_2}(\overline{v}; r''') \subset B^{\mathcal{C}}(\overline{v}; r)$, C_1 being the constant appearing in (5.14) and r being the radius appearing in (5.58). As $\|\dot{\Phi}(V)\|_Y \le C\|\dot{\Phi}(V)\|_Z$, we verify validity of (2.14).

(II) *Two-Dimensional Case.* We have to notice that it holds true that $H^s(\Omega) \subset \mathcal{C}(\overline{\Omega})$ only for $s > 1$ but not for $s = 1$, which means that $Z \subset Y$ fails now. So, let $1 < s < 2$. Then, since $\|v\|_{H^s} \le C\|v\|_{H^2}^{\alpha}\|v\|_{L_2}^{1-\alpha}$ for $v \in H^2(\Omega)$ with $\alpha = \frac{s}{2}$, it follows that $\|v\|_{\mathcal{C}} \le C\|v\|_{H^2}^{\alpha}\|v\|_{L_2}^{1-\alpha}$, $v \in H^2(\Omega)$. Therefore, there exists $r''' > 0$ such that $\overline{B}^{H^2}(0; C_2) \cap B^{L_2}(\overline{v}; r''') \subset B^{\mathcal{C}}(\overline{v}; r)$, C_2 being the constant appearing in (5.15) and r being the radius appearing in (5.58).

Meanwhile, set an additional space

$$W = \left\{ \begin{pmatrix} v \\ \rho \end{pmatrix}; \ v \in H_N^2(\Omega) \cap H_m^1(\Omega) \ \text{and} \ \rho \in H_N^2(\Omega) \right\}.$$

Since $\|w\|_{H^s} \leq C\|w\|_{H^2}^{\beta}\|w\|_{H^1}^{1-\beta}$ for $w \in H^2(\Omega)$ with $\beta = s - 1$, it follows that $\|\dot{\Phi}(V)\|_Y \leq C\|\dot{\Phi}(V)\|_W^{\beta}\|\dot{\Phi}(V)\|_Z^{1-\beta}$ for $V \in W$. Therefore, we observe from (5.49) that $\|\dot{\Phi}(V)\|_Y \leq C\|\dot{\Phi}(V)\|_Z^{1-\beta}$ for $V \in \overline{B}^W(0; C_2)$. This together with (5.58) then yields that

$$C\|\dot{\Phi}(V)\|_Z \geq |\Phi(V) - \Phi(\overline{V})|^{1-\theta'}, \qquad V \in \overline{B}^W(0; C_2) \cap B^X(\overline{V}; r'''),$$

with $\theta' = 1 - \frac{1-\theta}{1-\beta}$. Thereby, if s is taken so that $1 < s < 1 + \theta$, then θ' can be a positive exponent, that is, because of (5.15), (2.14) is fulfilled.

(III) *Three-Dimensional Case.* Taking account of $H^s(\Omega) \subset \mathcal{C}(\overline{\Omega})$ only for $s > \frac{3}{2}$, fix an exponent s so that $s > \frac{3}{2}$. On the other hand, let m denote an integer such that $s < 2(m + 1)$. Then, since $H^s(\Omega) \subset \mathcal{C}(\overline{\Omega})$ and since $\|v\|_{H^s} \leq C\|v\|_{H^{2(m+1)}}^{\alpha}\|v\|_{L_2}^{1-\alpha}$ for $v \in H^{2(m+1)}(\Omega)$ with $\alpha = \frac{s}{2(m+1)}$, it follows that $\|v\|_{\mathcal{C}} \leq C\|v\|_{H^{2(m+1)}}^{\alpha}\|v\|_{L_2}^{1-\alpha}$, $u \in H^{2(m+1)}(\Omega)$. Therefore, there exists $r''' > 0$ such that $\overline{B}^{H^{2(m+1)}}(0; C_{3,m}) \cap B^{L_2}(\overline{v}; r''') \subset B^{\mathcal{C}}(\overline{v}; r)$, $C_{3,m}$ being the constant appearing in (5.18) and r being the radius appearing in (5.58).

Meanwhile, set an additional space

$$W = \left\{ \begin{pmatrix} v \\ \rho \end{pmatrix}; \ v \in H_N^{2(m+1)}(\Omega) \cap H_m^1(\Omega) \ \text{and} \ \rho \in H_N^2(\Omega) \right\}.$$

Since $\|w\|_{H^s} \leq C\|w\|_{H^{2(m+1)}}^{\beta}\|w\|_{H^1}^{1-\beta}$ for $u \in H^{2(m+1)}(\Omega)$ with $\beta = \frac{s-1}{2m+1}$, it follows that $\|\dot{\Phi}(V)\|_Y \leq C\|\dot{\Phi}(V)\|_W^{\beta}\|\dot{\Phi}(V)\|_Z^{1-\beta}$ for $V \in W$. Therefore, $\|\dot{\Phi}(V)\|_Y \leq C\|\dot{\Phi}(V)\|_Z^{1-\beta}$ for $V \in \overline{B}^W(0; C_{3,m})$. This together with (5.58) then yields that

$$C\|\dot{\Phi}(V)\|_Z \geq |\Phi(V) - \Phi(\overline{V})|^{1-\theta'}, \qquad V \in \overline{B}^W(0; C_{3,m}) \cap B^X(\overline{V}; r'''),$$

with $\theta' = 1 - \frac{1-\theta}{1-\beta}$. Thereby, if m is taken sufficiently large so that $m > \frac{s-\theta-1}{2\theta}$, then θ' can be a positive exponent, that is, because of (5.18), (2.14) is fulfilled.

5.7 Notes and Future Studies

There is a big gap in the assumption (5.2) for the regularity of domain Ω between the cases of dimension two and dimension three. In the three-dimensional case, the $\mathcal{C}^\infty$ regularity is not a necessary condition but only a sufficient condition. An accurate order of regularity which we need in the arguments is determined by the exponent θ appearing in (5.58). However, it is not possible to know precisely an optimal value of such θ for a given domain Ω.

Let us comment more on the assumption (5.13). In the one-dimensional case, for any U_0 taken as (5.10)–(5.11), (5.3) with an initial condition $U(0) = U_0$ possesses a global solution $U(t)$ and its component $u(t)$ always satisfies (5.13) ([OY01, (4.6)]). In the two-dimensional case, if $\|u_0\|_{L_1}$ is sufficiently small for $U_0 = {}^t(u_0, \rho_0)$, then it is guaranteed that (5.3) with $U(0) = U_0$ possesses a global solution $U(t)$ whose component $u(t)$ satisfies (5.13). For the details, see the proof of [Yag10, Proposition 12.2], Step 3. As remarked there, the norm $\|u(t)\|_{L_2}$ can be dominated by the quantity $N_{1,\log}(u(t))$, where $N_{1,\log}(u) = \int_\Omega u \log(u + 1) dx$, for any global solution. Thereby, in this case, it is possible to replace (5.13) by a weaker condition that

$$\sup_{0 \leq t < \infty} N_{1,\log}(u(t)) < \infty.$$

In the three-dimensional case, however, we know neither suitable sufficient conditions for U_0 which guarantee existence of global solution to (5.3) with $U(0) = U_0$ ([Wi10]) nor sufficient conditions for (5.13).

In the two-dimensional case, Feireisl–Laurençot–Petzeltová [FLP07] first proved the longtime convergence of global solutions by using the non-smooth version of the Łojasiewicz–Simon inequality due to Feireisl–Issard-Roch–Petzeltová [FIP04]. They treated classical solutions $u(x, t)$ and $\rho(x, t)$ to (5.1) in a smooth domain Ω satisfying $u(x, t) > 0$ in $\overline{\Omega}$ and global boundedness conditions

$$\sup_{0 \leq t < \infty} \|u(t)\|_{L_\infty} < \infty \quad \text{and} \quad \sup_{0 \leq t < \infty} \|\rho(t)\|_{L_\infty} < \infty.$$

As they did not use the Angle Condition, any order estimate of convergence like (5.52) was not obtained.

The results of this chapter for the one-dimensional case were already published by Iwasaki–Osaki–Yagi [IOY]. The results of the higher-dimensional cases are new.

Bibliography

[Ada75]   Adams, R.A.: Sobolev Spaces. Academic Press, San Diego (1975)
[AMY17]   Azizi, S., Mola, G., Yagi, A.: Longtime convergence for epitaxial growth model under Dirichlet conditions. Osaka J. Math. **54**, 689–706 (2017)
[AY17a]   Azizi, S., Yagi, A.: Dynamical system for epitaxial growth model under Dirichlet conditions. Sci. Math. Jpn. **80**, 109–122 (2017)
[AY17b]   Azizi, S., Yagi, A.: Homogeneous stationary solutions to epitaxial growth model under Dirichlet conditions. Sci. Math. Jpn. **80**, 123–131 (2017)
[Bre11]   Brezis, H.: Functional Analysis, Sobolev Spaces and Partial Differential Equations. Springer, Berlin (2011)
[Chi03]   Chill, R.: On the Łojasiewicz-Simon gradient inequality. J. Funct. Anal. **201**, 572–601 (2003)
[Chi06]   Chill, R.: The Lojasiewicz-Simon gradient inequality on Hilbert spaces. Proceedings of the Fifth European-Maghrebian Workshop on Semigroup Theory, Evolution Equations and Applications, pp. 25–36 (2006)
[CHJ09]   Chill, R., Haraux, A., Jendoubi, M.A.: Applications of the Łojasiewicz-Simon gradient inequality to gradient-like evolution equations. Anal. Appl. **7**, 351–372 (2009)
[DL84a]   Dautray, R., Lions, J.L.: Analyse mathématique et calcul numérique pour les sciences et les techniques, vol. 1. Masson, Paris (1984)
[DL84b]   Dautray, R., Lions, J.L.: Analyse mathématique et calcul numérique pour les sciences et les techniques, vol. 2. Masson, Paris (1984)
[Die60]   Dieudonné, J.: Foundations of Modern Analysis. Academic Press, San Diego (1960)
[EH66]   Ehrlich, G., Hudda, F.G.: Atomic view of surface self-diffusion: tungsten on tungsten. J. Chem. Phys. **44**, 1039–1049 (1966)
[FIP04]   Feireisl, E., Issard-Roch, F., Petzeltová, H.: A non-smooth version of the Lojasiewicz-Simon theorem with applications to non-local phase-field systems. J. Differ. Equ. **199**, 1–21 (2004)
[FLP07]   Feireisl, E., Laurençot, P., Petzeltová, H.: On convergence to equilibria for the Keller-Segel chemotaxis model. J. Differ. Equ. **236**, 551–569 (2007)
[GMY]   Grasselli, M., Mola, G., Yagi, A.: On the longtime behavior of solutions to a model for epitaxial. Osaka J. Math. **48**, 987–1004 (2011)
[Gri85]   Grisvard, P.: Elliptic Problems in Nonsmooth Domains. Pitman, London (1985)
[HV96]   Herrero, M.A., Velázquez, J.J.L.: Chemotactic collapse for the Keller-Segel model. J. Math. Biol. **35**, 177–194 (1996)
[HV97]   Herrero, M.A., Velázquez, J.J.L.: A blow-up mechanism for a chemotaxis model. Ann. Sc. Norm. Super. Pisa Cl. Sci. **24**, 633–683 (1997)

[Hör90]   Hörmander, L.: An Introduction to Complex Analysis in Several Variables, 3rd revised
           edn. North-Holland, Amsterdam (1990)
[HW01]   Horstmann, D., Wang, G.: Blow-up in a chemotaxis model without symmetry assump-
           tions. Euro. J. Appl. Math. **12**, 159–177 (2001)
[IOY]   Iwasaki, S., Osaki, K., Yagi, A.: Asymptotic convergence of solutions for one-
           dimensional Keller-Segel equations. http://hdl.handle.net/11094/77680
[IY]   Iwasaki, S., Yagi, A.: Asymptotic convergence of solutions for Laplace reaction-
           diffusion equations. Nonlinear Analysis RWA (to appear)
[Jen98]   Jendoubi, M.A.: A simple unified approach to some convergence theorem by L. Simon.
           J. Funct. Anal. **153**, 187–202 (1998)
[JOHGSSO94] Johnson, M.D., Orme, C., Hunt, A.W., Graff, D., Sudijono, J., Sauder, L.M., Orr,
           B.G.: Stable and unstable growth in molecular beam epitaxy. Phys. Rev. Lett. **72**, 116–
           119 (1994)
[LM68]   Lions, J.L., Magenes, E.: Problème aux limites non homogènes et applications, vol. 1
           Dunord, Paris (1968)
[Loj63]   Łojasiewicz, S.: Une propriété topologique des sous-ensembles analytique réels. Collo-
           ques internationaux du C.N.R.S.: Les équations aux dérivées partielles, Paris, 1962, pp.
           87–89. Editions du C.N.R.S., Paris (1963)
[Loj65]   Łojasiewicz, S.: Ensembles Semi-Analytiques. Publ. Inst. Hautes Etudes Sci., Bures-
           sur-Yvette (1965)
[LZ99]   Łojasiewicz, S., Zurro, M.A.: On the gradient inequality. Bull. Polish Acad. Sci. Math.
           47, 143–145 (1999)
[Mat78]   Matano, H.: Convergence of solutions of one-dimensional semilinear parabolic equa-
           tions. J. Math. Kyoto Univ. **18**, 221–227 (1978)
[Mul57]   Mullins, W.W.: Theory of thermal grooving. J. Appl. Phys. **28**, 333–339 (1957)
[OY01]   Osaki, K., Yagi, A.: Finite dimensional attractor for one-dimensional Keller-Segel
           equations. Funkc. Ekvac. **44**, 441–469 (2001)
[PR96]   Poláčik, P., Rybakowski, K.P.: Nonconvergent bounded trajectories in semilinear heat
           equations. J. Differ. Equ. **124**, 472–494 (1996)
[PS02]   Poláčik, P., Simondon, F.: Nonconvergent bounded solutions of semilinear heat equa-
           tions on arbitrary domains. J. Differ. Equ. **186**, 586–610 (2002)
[RH]   Rybka, P., Hoffmann, K.-H.: Convergence of solutions to Cahn-Hilliard equation.
           Commun. Partial Differ. Equ. **24**(5&6), 1055–1077 (1999)
[SS66]   Schwoebel, R.L., Shipsey, E.J.: Step motion on crystal surfaces. J. Appl. Phys. **37**,
           3682–3686 (1966)
[Tan75]   Tanabe, H.: Equations of Evolution. Iwanami Shoten, Tokyo (1975). (in Japanese)
           English translation: Pitman, London (1979)
[Tan97]   Tanabe, H.: Functional Analytic Methods for Partial Differential Equations. Marcel
           Dekker, New York (1997)
[Tri78]   Triebel, H.: Interpolation Theory, Function Spaces, Differential Operators. North-
           Holland, Amsterdam (1978)
[Wi10]   Winkler, M.: Aggregation vs. global diffusive behavior in the higher-dimensional
           Keller-Segel model. J. Differ. Equ. **248**, 2889–2905 (2010)
[Yag10]   Yagi, A.: Abstract Parabolic Evolution Equations and their Applications. Springer,
           Berlin (2010)
[Yag17]   Yagi, A.: Real sectorial operators. Bull. South Ural State Univ. Ser. MMCS **10**, 97–112
           (2017)
[Yag]   Yagi, A.: Abstract Parabolic Evolution Equations and the Łojasiewicz-Simon Inequal-
           ity, vol. 1, Abstract Theory. Springer, Berlin (2021)

[Yos80]   Yosida, K.: Functional Analysis, 6th edn. Springer, Berlin (1980)
[Zei88]   Zeidler, E.: Nonlinear Functional Analysis and Its Applications, vol. I. Springer, Berlin
           (1988)
[Zel68]   Zelenyak, T.I.: Stabilization of solutions of boundary values problems for a second-
           order parabolic equation with one space variable. Differ. Equ. **4**, 17–22 (1968).
           Translation from Differ. Uravn. **4**, 34–45 (1968)

Symbol Index

© The Author(s), under exclusive license to Springer Nature Singapore Pte Ltd. 2021   125
A. Yagi, *Abstract Parabolic Evolution Equations and Łojasiewicz–Simon Inequality II*,
SpringerBriefs in Mathematics, https://doi.org/10.1007/978-981-16-2663-0

Index

© The Author(s), under exclusive license to Springer Nature Singapore Pte Ltd. 2021   127
A. Yagi, *Abstract Parabolic Evolution Equations and Łojasiewicz–Simon Inequality II*,
SpringerBriefs in Mathematics, https://doi.org/10.1007/978-981-16-2663-0

Printed in the United States
by Baker & Taylor Publisher Services